physica status solidi (b)
basic research

Volumes 139 to 144

Subject and Author Index 1987

Volumes 99 to 104

AKADEMIE-VERLAG BERLIN
VCH PUBLISHERS INC. · NEW YORK/N.Y.

ISSN 0370-1972/0031-8965 phys. stat. sol. (b)/(a), Berlin **R20** (1987) 1—465

Substance Classification*)

S1. Transition metals and their alloys and compounds (for oxides, see S10)
- S1.1. Fe
- S1.2. Other 3d metals
- S1.3. Noble metals
- S1.4. Lanthanides
- S1.5. Actinides
- S1.61. Carbides, silicides, germanides
- S1.62. Nitrides, phosphides, arsenides
- S1.63. Sulfides, selenides, tellurides

S2. Alkali metals and their alloys and compounds (for alkali halides, see S9.11)

S3. Alkaline earth metals and their alloys and compounds (for alkaline earth halides, see S9.12)

S4. Other metals and their alloys

S5. Nonmetallic elements
- S5.11. Silicon
- S5.12. Germanium
- S5.2. Solidified Gases

S6. Group IV compounds (for transition metal compounds, see S1.61)

S7. Group V compounds (for transition metal compounds, see S1.62)
- S7.1. III–V compounds
- S7.11. Phosphides
- S7.12. Arsenides
- S7.13. Antimonides
- S7.15. Mixed crystals
- S7.16. Higher compounds

S8. Chalcogenides (for transition metal compounds, see S1.63)
- S8.1. II–VI compounds
- S8.11. Sulfides
- S8.12. Selenides
- S8.13. Tellurides
- S8.15. Mixed crystals
- S8.16. Higher compounds

S9. Halides
- S9.11. Alkali halides
- S9.12. Alkaline earth halides
- S9.15. Mixed crystals
- S9.16. Higher compounds

S10. Oxides
- S10.1. Simple oxides
- S10.15. Mixed crystals

S11. Other higher compounds
- S11.1. Salts of oxy-acids
- S11.2. Spinels, garnets, ferrites

S12. Organic compounds
- S12.1. Quasi-one-dimensional systems

For comments on the use of the classification scheme see p. 10.

---

*) Outside the ICSTI Classification for Physics.

physica status solidi (b)

basic research

physica status solidi (a)

applied research

SUBJECT AND AUTHOR INDEX

Volumes 139 to 144, 1987 Volumes 99 to 104, 1987

ISSN 0370-1972/0031-8965

Edited by
E. BORCHARDT

AKADEMIE-VERLAG · BERLIN

and

VCH PUBLISHERS INC. · NEW YORK/N.Y.

U.S. Postmaster: Send address changes to physica status solidi (b) and physica status solidi (a),
VCH Publishers, Inc., 303 NW 12th Avenue, Deerfield Beach,
FL 33442-1788, USA.

Enquiries, subscriptions, and orders for this Subject and Author Index
or single copies of physica status solidi (b) and physica status solidi (a) from customers in the USA and CANADA should be directed to:

VCH Publishers, Inc., 303 NW 12th Avenue, Deerfield Beach, FL 33442—1788, USA;

from all other countries to:

Akademie-Verlag Berlin, Leipziger Str. 3—4, DDR-1086 Berlin, GDR.

Schriftleiter und verantwortlich für den Inhalt der Zeitschriften:
Professor Dr. E. Gutsche. Leipziger Str. 3—4, Postfach 1233, DDR-1086 Berlin.

Herausgeber und verantwortlich für den Inhalt des Registers:
Dipl.-Phys. E. Borchardt, Leipziger Str. 3—4, Postfach 1233, DDR-1086 Berlin.

Verlag:
Akademie-Verlag Berlin, Leipziger Str. 3—4, DDR-1086 Berlin, Fernruf: 223 622 1 und 223 6229; Telex-Nr.: 114420;
Bank: Staatsbank der DDR, Berlin, Kto.-Nr.: 6836-26-20712.

Veröffentlicht unter der Lizenznummer 1310 und 1620 des Presseamtes beim Vorsitzenden des Ministerrates der Deutschen Demokratischen Republik.

Gesamtherstellung:
VEB Druckerei „Thomas Müntzer", DDR-5820 Bad Langensalza.

Bestellnummer des Registers: 1068/85 R 20.

Printed in the German Democratic Republic.

AN (EDV) 20831/20735

Preface

The twentieth index volume contains volumes 139 to 144 of "physica status solidi (b) — basic research" as well as volumes 99 to 104 of "physica status solidi (a) — applied research", i.e. all volumes having appeared in 1987 containing 1287 papers.

Beginning from January 1986 the International Classification System for Physics (Condensed Matter, items 60 and 70) of the International Council for Scientific and Technical Information, Abstracting Board completed by item 85 and a Substance Classification (both outside the ICSTI Classification for Physics) will be used for quoting papers in physica status solidi. Nevertheless, before taking over these subject classifications into the index volume they were once more carefully revised and in some cases improved and completed in order to get a most uniform and clear classification for all papers. The average number of quotations per paper in this index volume is about 2.7.

Again I have to express my thanks to the readers for their interest, and I will always readily take up any critical remarks and proposals for improvements. I am especially obliged to Prof. Dr. E. Gutsche, Dr. H.-J. Hänsch, Dr. H. Lange, Dr. S. Oberländer, and Dipl. Phys. K. Müller for giving me advice and unselfish help. Furthermore I have to thank Mrs. E. Schwan for her exact and reliable secretary's work at the manuscript.

Berlin, June 1988

Eva Borchardt

International Classification System for Physics*)

60. Condensed matter: structure, mechanical and thermal properties
61. Structure of liquids and solids; crystallography
    (see also 68.20. Solid surface structures, 71. Electron states)
    61.10. X-ray determination of structures
        (for specific determinations, see 61.55 to 61.80)
    61.12. Neutron determination of structures
        (for specific determinations, see 61.55 to 61.80)
    61.14. Electron determination of structures
        (for specific determinations, see 61.55 to 61.80)
    61.16. Other determination of structures
        (for specific determinations, see 61.55 to 61.80)
    61.20. Classical, semiclassical, and quantum theories of liquid structure
        (for electronic states, see 71; for liquid helium, see 67)
    61.25. Studies of specific liquid structures
    61.30. Liquid crystals
    61.40. Amorphous and polymeric materials
    61.50. Crystalline state (including molecular motions in solids)
        (for magnetic structure and spin systems, see 75.25)
    61.55. Specific structures of elements and alloys
    61.60. Specific structures of inorganic compounds
    61.65. Specific structures of organic compounds
    61.70. Defects in crystals
        (see also 61.80. Radiation damage, 62. Mechanical and acoustical properties, 71.55. Impurities and defect levels, 76.30. EPR of color centers and other defects, 78.50. Impurity and defect absorption in solids)
    61.80. Radiation damage and other irradition effects
        (for techniques of structure determination, see 61.10 to 61.16; for electron and ion impact phenomena, see 79.20)
    61.90. Other topics in structure of liquids and solids

62. Mechanical and acoustical properties of condensed matter
    (see also 61.70. Defects in crystals, 68.30. Surfaces and interfaces)
    62.10. Mechanical properties of liquids
        (for viscosity of liquids, see also 66.20)
    62.20. Mechanical properties of solids (related to microscopic structure)
    62.30. Mechanical and elastic waves
    62.40. Anelasticity, internal friction, and mechanical resonances
        For thermomechanical effects, see 65.70
        For magnetomechanical effects, see 75.80
        For piezoelectric effects, see 77.60
        For elasto-optical effects, see 78.20
    62.50. High-pressure and shock-wave effects in solids
    62.60. Acoustical properties of liquids
        For lattice dynamics, phonons, see 63
        For second sound in quantum fluids, see 67.40
    62.65. Acoustical properties of solids
        For magnetoacoustic effects in metals, see 75.80
        For acoustoelectric effects, see 72.50
        For acousto-optical effects, see 78.20
    62.80. Ultrasonic relaxation
        (see also 74.30. Ultrasonic attenuation in superconductors)
    62.90. Other topics in mechanical and acoustical properties of condensed matter

*) Excerpt, reproduced with permission of International Council for Scientific and Technical Information (ICSTI).

63. Lattice dynamics and crystal statistics
    (see also 65. Thermal properties, 66.70. Thermal conduction, 68.30. Dynamics of surface and interface vibrations, 78.30. Infrared and Raman spectra)
    63.10. General theory
    63.20. Phonons and vibrations in crystal lattices
    63.50. Vibrational states in disordered systems
    63.70. Statistical mechanics of lattice vibrations
        (see also 65. Thermal properties of condensed matter, and 66.70. Thermal conduction)
    63.75. Statistical mechanics of displacive phase-transitions
    For order–disorder and statistical mechanics of model systems, see 64.60
    For crystallographic aspects of polymorphic and order–disorder transformations, see 61.50
    63.90. Other topics in lattice dynamics and crystal statistics

64. Equations of state, phase equilibria, and phase transitions
    64.10. General theory of equations of state and phase equilibria
    64.30. Equations of state of specific substances
        (see also 65.70. Thermal expansion)
    64.60. General studies of phase transitions
        (for magnetic, superconducting, and quantum fluid critical phenomena, see 75, 74, and 67, respectively)
    64.70. Phase equilibria, phase transitions, and critical points of specific substances
    64.75. Solubility, segregation, and mixing
    64.80. Other phase properties of systems
    64.90. Other topics in equations of state, phase equilibria, and phase transitions

65. Thermal properties of condensed matter
    (see also 63. Lattice dynamics; for thermodynamic properties of quantum fluids, see 67.40; for thermal properties of solid helium, see 67.80)
    65.20. Heat capacities of liquids
    65.40. Heat capacities of solids
        (for specific heat of superconductors, see 74.30; for specific heat of magnetic systems, see 75.40)
    65.50. Thermodynamic properties and entropy
    65.70. Thermal expansion and thermomechanical effects
        (see also 64.30. Equations of state)
    For thermal conduction in nonmetallic liquids, see 66.60; for nonmetallic solids, see 66.70
    For electronic thermal conduction, see 72.10 and 72.20
    For thermal conductivity of superconductors, see 74.30
    For pyroelectric and electrocaloric effects, see 77.70
    65.90. Other topics in thermal properties of condensed matter

66. Transport properties of condensed matter (nonelectronic)
    66.10. Diffusion and ionic conduction in liquids
    66.20. Diffusive momentum transport
        (for viscosity of liquids, see also 62.10)
    66.30. Diffusion in solids
    66.60. Thermal conduction in nonmetallic liquids
        (for thermal conduction in liquid metals, see 72.15)
    66.70. Nonelectronic thermal conduction and heat-pulse propagation in nonmetallic solids
        (for thermal conduction in solid metals, see 72.15; for statistical mechanics of lattice vibrations, see 63.70)
    66.90. Other topics in nonelectronic transport properties

67. Quantum fluids and solids; liquid and solid helium
    - 67.20. Quantum effects on the structure and dynamics of nondegenerate fluids
    - 67.40. Boson degeneracy and superfluidity of helium-4
    - 67.50. Fermi fluids; liquid helium-3
    - 67.60. Mixed systems; liquid helium-3, -4 mixtures
    - 67.70. Films (including physical adsorption)
    - 67.80. Solid helium and related quantum crystals
    - 67.90. Other topics in quantum fluids (e.g., neutron-star matter)

68. Surfaces and interfaces; thin films and whiskers (for impact phenomena, see 79; for crystals growth, see 61.50)
    - 68.10. Fluid surfaces and fluid–fluid interfaces
    - 68.15. Liquid thin films
    - 68.20. Solid surface structures
    - 68.25. Mechanical and acoustical properties of solid surfaces and interfaces
    - 68.30. Dynamics of solid surfaces and interface vibrations
    - 68.40. Surface energy of solids; thermodynamic properties
    - 68.45. Solid–fluid interface processes
    - 68.48. Solid–solid interfaces (including bicrystals) (for grain boundaries, see 61.70)
    - 68.55. Thin film growth, structure, and epitaxy
    - 68.60. Physical properties of thin films: nonelectronic
    - 68.70. Whiskers and dendrites: growth, structure, and nonelectronic properties
    - 68.90. Other topics in the structure and nonelectronic properties of surfaces and thin films

70. Condensed matter: electronic structure; electrical, magnetic, and optical properties

71. Electronic states (see also 63. Lattice dynamics, 73. Electronic structure and electrical properties of surfaces, interfaces, and thin films)
    - 71.10. General theories and computational techniques
    - 71.20. Electronic density of states determinations (including energy states of liquid semiconductors) (see also 65.40. Electronic heat capacity)
    - 71.25. Nonlocalized single-particle electronic states
    - 71.30. Metal–insulator transitions
    - 71.35. Excitons and related phenomena (including electron–hole drops)
    - 71.36. Polaritons (including photon–phonon and photon–magnon interactions)
    - 71.38. Polarons and electron–phonon interactions (see also 63.20. phonon–electron interactions in lattices)
    - 71.45. Collective effects
    - 71.50. Localized single-particle electronic states (excluding impurities)
    - 71.55. Impurity and defect levels
    - 71.60. Positron states (see also 78.70. Positron annihilation)
    - 71.70. Level splitting and interactions (see also 75.10. — in magnetic phenomena, 75.30. Exchange and superexchange interactions, 73.20. Electronic surface states)
    - 71.90. Other topics in electron states

72. Electronic transport in condensed matter
    (for surfaces, interfaces, and thin films, see 73)
    72.10. Theory of electronic transport; scattering mechanism
    72.15. Electronic conduction in metals and alloys
    72.20. Conductivity phenomena in semiconductors and insulators
        (for nonelectronic conduction, see 66.70)
    72.30. High-frequency effects; plasma effects
    72.40. Photoconduction and photovoltaic effects; photodielectric effects
    72.50. Acoustoelectric effects
    72.55. Magnetoacoustic effects
    72.60. Mixed conductivity and conductivity transitions
    72.70. Noise processes and phenomena
    72.80. Conductivity of specific semiconductors and insulators
    72.90. Other topics in electronic transport in condensed matter

73. Electronic structure and electrical properties of surfaces, interfaces, and thin films
    73.20. Electronic surface states
        (for emission and impact phenomena, see 79)
    73.25. Surface conductivity
    73.30. Surface double layers, Schottky barriers, and work functions
    73.40. Interfaces
    73.60. Electronic properties of thin films
    73.90. Other topics in electrical properties of surfaces, interfaces, and thin films

74. Superconductivity
    74.10. Occurrence, critical temperature
    74.20. Theory
    74.30. General properties
    74.40. Fluctuations and critical effects
    74.50. Tunneling phenomena, Josephson effect, and proximity effects
    74.55. Type-I superconductivity
    74.60. Type-II superconductivity
    74.70. Superconducting materials
    74.90. Other topics in superconductivity

75. Magnetic properties and materials
    75.10. General theory and models of magnetic ordering
        (see also 71.25. Nonlocalized single-particle electronic states, 71.70. Level splitting and interactions)
    75.20. Diamagnestism and paramagnetism
    75.25. Spin arrangements in magnetically ordered materials
        (neutron studies, etc.)
    75.30. Magnetically ordered materials, other intrinsic properties
        (for critical point effects, see 75.40)
    75.40. Critical-point effects, specific heats, short-range order
        (see also 65.40. Heat capacities)
    75.50. Studies of specific magnetic materials
    75.60. Domain effects, magnetization curves, and hysteresis
    75.70. Magnetic films and plates
    75.80. Magnetomechanical and magnetoelectric effects, magnetostriction
      For galvanomagnetic effects, see 72.15 and 72.20
      For magneto-optical effects, see 78.20
    75.90. Other topics in magnetic properties and materials

76. Magnetic resonances and relaxation in condensed matter; Mössbauer effect
    76.20. General theory of resonances and relaxation
    76.30. Electron paramagnetic resonance and relaxation
    76.40. Diamagnetic and cyclotron resonances
    76.50. Ferromagnetic, antiferromagnetic, and ferrimagnetic resonances; spin-wave resonance (see also 75.30. Spin waves)
    76.60. Nuclear magnetic resonance and relaxation
    76.70. Magnetic double resonances and cross effects
    76.80. Mössbauer effect; other γ-ray spectroscopy
    76.90. Other topics in magnetic resonances and relaxation

77. Dielectric properties and materials
    (for conductivity phenomena, see 72.20 and 72.80)
    77.20. Permittivity
    77.30. Polarization and depolarization effects
    77.40. Dielectric loss and relaxation
    77.50. Dielectric breakdown and space-charge effects
    77.55. Dielectric thin films
    77.60. Piezoelectricity and electrostriction (for piezo-optical effects, see 78.20)
    77.70. Pyroelectric and electrocaloric effects
    77.80. Ferroelectricity and antiferroelectricity
    77.85. Electrical resonances
    77.90. Other topics in dielectric properties and materials

78. Optical properties and condensed-matter spectroscopy and other interactions of matter with particles and radiation
    (for phonon spectra, see 63)
    78.20. Optical properties and materials
    78.30. Infrared and Raman spectra and scattering
    78.35. Brillouin and Rayleigh scattering
    78.40. Visible and ultraviolet spectra
    78.45. Stimulated emission
    78.50. Impurity and defect absorption in solids
    78.55. Photoluminescence
    78.60. Other luminescence and radiative recombination
    For photoconduction and photovoltaic effects, see 72.40
    78.65. Optical properties of thin films
    78.70. Other interactions of matter with particles and radiation
    78.90. Other topics in optical properties of condensed matter and other interactions of matter with particles and radiation

79. Electron and ion emission by liquids and solids; impact phenomena
    79.20. Impact phenomena, including electron spectra and sputtering
    79.40. Thermionic emission
    79.60. Photoemission and photoelectron spectra
    79.70. Field emission and field ionization
    79.75. Exoelectron emission
    79.80. Resonance tunneling
    79.90. Other topics in emission and impact phenomena in condensed matter

85. Devices**)

**) Outside the ICSTI Classification for Physics.

Substance Classification (according to physica status solidi)

S1. Transition metals and their alloys and compounds
(for oxides, see S10)
- S1.1. Fe
- S1.2. Other 3d metals
- S1.3. Noble metals
- S1.4. Lanthanides
- S1.5. Actinides
- S1.61. Carbides, silicides, germanides
- S1.62. Nitrides, phosphides, arsenides
- S1.63. Sulfides, selenides, tellurides

S2. Alkali metals and their alloys and compounds
(for alkali halides, see S9.11)

S3. Alkaline earth metals and their alloys and compounds
(for alkaline earth halides, see S9.12)

S4. Other metals and their alloys

S5. Nonmetallic elements
- S5.11. Silicon
- S5.12. Germanium
- S5.2. Solidified Gases

S6. Group IV compounds
(for transition metal compounds, see S1.61)

S7. Group V compounds
(for transition metal compounds, see S1.62)
- S7.1. III–V compounds
- S7.11. Phosphides
- S7.12. Arsenides
- S7.13. Antimonides
- S7.15. Mixed crystals
- S7.16. Higher compounds

S8. Chalcogenides
(for transition metal compounds, see S1.63)
- S8.1. II–VI compounds
- S8.11. Sulfides
- S8.12. Selenides
- S8.13. Tellurides
- S8.15. Mixed crystals
- S8.16. Higher compounds

S9. Halides
- S9.11. Alkali halides
- S9.12. Alkaline earth halides
- S9.15. Mixed crystals
- S9.16. Higher compounds

S10. Oxides
- S10.1. Simple oxides
- S10.15. Mixed crystals

S11. Other higher compounds
- S11.1. Salts of oxy-acids
- S11.2. Spinels, garnets, ferrites

S12. Organic compounds
- S12.1. Quasi-one-dimensional systems

For comments on the use of the classification scheme see p. 10.

Some Comments on the Use of the Subject and Author Index

The "International Classification System for Physics" (the condensed matter sections 60 and 70) completed by a "Substance Classification" (according to physica status solidi) is used. On cover two the second-level headings of the condensed matter sections and on pages 4 to 8 the third-level headings are reproduced. As a help for using the subject index on pages 11 and 12 in addition also a series of important fourth-level headings is given. The substance classification is published on cover three and on page 9.

All papers will be quoted under each relevant section. There is made a distinction between one or two principal entries (refering to the main subject matter of the paper) and secondary entries (refering to subsidiary subject matter and substance, respectively).In the principal entries the title of the paper is given in its original language and in English. In addition, every principal entry is marked by * in front of the title of the paper. In the secondary entries the title is only given in English.

On the line following the title the subject classification is given (in parantheses) in the same manner as in the volumes. Each entry gives the volume, the number of the issue, the initial and final page numbers, and the year in which the paper appeared. A paper of physica status solidi (b) or physica status solidi (a) is marked by (b) or (a) in front of the line, respectively. The letter "K" before the page numbers indicates a short note published in the offset part of the journal.

Review articles appear, at the head of the list of entries under a given subject and are marked. These articles are listed in the chronological order of their publication, irrespective of being a paper of physica status solidi (b) or physica status solidi (a). The subject classification for a review article consisting of several parts is based on the article as a whole rather than on the individual parts.

The entries for original papers, short notes, and errata are made together in a single list for each subject and follow the entries for review articles. These are again ordered according to data of publication, irrespective of being a paper of physica status solidi (b) or physica status solidi (a). Original papers are sometimes published in several parts under the same principal title (in general with special subtitles). The various parts of papers of this type are treated separately and each part is classified according to its particular contents.

In quoting errata the following rule is obeyed. If the erratum refers to a paper quoted in the preceding index, it is listed in the chronological order of its publication otherwise the erratum is quoted together with the paper to which it refers.

The author index gives the names of the authors in alphabetical order. Each name is followed by the bibliographic details of the paper. A paper of physica status solidi (b) or physica status solidi (a) is again marked by (b) or (a) in front of the number of the volume, respectively. In the last column the subject classification of the paper is presented in the same manner as in the volumes. When several papers have been published by the same author they are arranged according to the dates of publication, irrespective of being a paper of physica status solidi (b) or physica status solidi (a). To find a paper of your interest please choose first the special entry of the subject index and then use the volume number and page.

In using the subject index attention should be given to the following details:

61.10 to 61.16 Theory and experimental techniques of determination of structures
- for specific determinations see 61.55 to 61.80

61.16 Other determination of structures
- electron microscopy and field-ion microscopy determinations
- EPR and NMR determinations

61.25 Studies of specific liquid structures
- macromolecular and polymer solutions
- molten salts
- liquid metals

61.40 Amorphous and polymeric materials
- glasses, polymers, elastomers, plastics

61.50 Crystalline state (including molecular motions in solids)
- physics of crystal growth
- crystal symmetry: models, space groups, crystalline systems and classes
- crystal morphology and orientation
- crystallographic aspects of polymorphic and order–disorder transformations

61.70 Defects in crystals
- grain and twin boundaries
- stacking faults and other planar or extended defects
- doping and implantation of impurities
- impurity concentration, distribution, and gradients

61.80 Radiation damage and other irradiation effects
- channeling, blocking, energy loss of particles
- X-rays, γ-rays, electrons and positrons, neutrons, atoms and molecules, ions
- for ion implantation see 61.70

63.20 Phonons and vibrations in crystal lattices
- phonon–phonon, phonon–electron, and phonon–defect interactions

64.60 General studies of phase transitions
- order–disorder and statistical mechanics of model systems
- equilibrium properties near single critical points, critical exponents
- dynamic critical phenomena
- multicritical points
- metastable phases

64.80 Other phase properties of systems
- stoichiometry and homogeneity
- microstructure

66.30 Diffusion in solids
- theory of diffusion and ionic conduction in solids
- self-diffusion and ionic conduction in non-metals
- diffusion, migration, and displacement of impurities and other defects

71.25 Nonlocalized single-particle electronic states
- techniques of band-structure calculation
- measurement of Fermi surface parameters (including dHvA, magnetoacoustic, positron annihilation, and cyclotron resonance studies, etc.)
- effective mass and *g*-factors
- band structure of crystalline metals, semiconductors, and insulators

71.45 Collective effects
- exchange, correlation, dielectric and magnetic functions, plasmons
- Fermi-Thomas models
- calculations of total electronic binding energy

71.70 Level splitting and interactions
- crystal and ligand fields
- spin–orbit coupling, Zeeman, Stark, and strain splitting
- exchange interactions
- nuclear states and interactions

72.10 Theory of electronic transport; scattering mechanism
- general formulation of transport theory
- scattering by phonons, magnons, and other nonlocalized excitations
- scattering by point defects, dislocations, surfaces, and other imperfections (including Kondo effect)

72.15 Electronic conduction in metals and alloys
- galvanomagnetic and other magnetotransport effects
- thermomagnetic and thermoelectric effects
- collective modes
- scattering mechanism and Kondo effect

72.20 Conductivity phenomena in semiconductors and insulators
- scattering mechanism
- piezoresistance
- galvanomagnetic and other magnetotransport effects
- thermomagnetic and thermoelectric effects

73.40 Interfaces
- contact resistance, contact potential, rectification
- tunneling: general
- p–n junctions
- MIS, MIM, MSM, SIS, and SMS structures

75.30 Magnetically ordered materials, other intrinsic properties
- saturation moments and magnetic susceptibility
- spin waves
- exchange and superexchange interactions
- anisotropy
- magnetic impurity interactions
- magnetic phase boundaries
- magnetocaloric effect

75.50 Studies of specific magnetic materials
- ferromagnetism
- antiferromagnetics, ferrimagnetics
- amorphous magnetic materials
- magnetic liquids

75.60 Domain effects, magnetization curves, and hysteresis
- high coercitivity materials
- fine-particle systems
- magnetic aftereffects
- magnetic annealing and temperature-hysteresis effects

78.20 Optical properties and materials
- general theory
- optical constants and parameters
- optical rotatory power
- birefringence
- piezo-, elasto-, acousto-, electro-, magneto-, and thermo-optical effects

78.60 Other luminescence and radiative recombination
- electroluminescence
- cathodo- and iono-luminescence
- thermoluminescence
- sono- and tribo-luminescence
- chemiluminescence

78.70 Other interactions of matter with particles and radiation
- positron annihilation
- X-ray scattering, absorption, and emission
- microwave and radiofrequency (including resonances)

79.20 Impact phenomena, including electron spectra and sputtering
- electron impact: Auger and secondary emission
- atom, molecule, and ion impact
- atomic and molecular beam interactions

Subject Index

60. Condensed matter : structure, mechanical and thermal properties

(see also 68.20. Solid surface structures, 71. Electron states)

61.10. X-ray determination of structures
(for specific determinations, see 61.55 to 61.80)

P. H. BEZIRGANYAN and V. S. HAROUTYUNYAN
*On the Interference Effect Arising in the Bicrystal X-Ray Interferometer with Wedge-Shaped Gap. I. Plane Wave Theory
(61.10)
(a) Vol. **99**, No. 1, p. 37—46 (1987).

H. LÖFFLER, G. WENDROCK, and K. PAWELZYK
The Influence of Small Mg Additions on the Stability of Guinier-Preston Zones in Al–4.5 at%–Zn–x Mg Alloys
(61.55; 61.10; 64.70; S4)
(a) Vol. **99**, No. 1, p. 73—80 (1987).

V. S. HAROUTYUNYAN and P. H. BEZIRGANYAN
*On the Interference Effect Arising in the Bicrystal X-Ray Interferometer with Wedge-Shaped Gap. II. Spherical Wave Theory
(61.10)
(a) Vol. **99**, No. 2, p. 329—338 (1987).

J. BĄK-MISIUK, J. GRONKOWSKI, J. HÄRTWIG, and W. WIERZCHOWSKI
*X-Ray Spherical-Wave Diffraction in the Bragg Case for Bent Si Crystals
(61.10; 61.55; S5.11; S10.1)
(a) Vol. **99**, No. 2, p. 345—351 (1987).

V. HOLÝ and K. T. GABRIELYAN
*Dyson and Bethe-Salpeter Equations for Dynamical X-Ray Diffraction in Crystals with Randomly Placed Defects
(61.10; 61.70)
(b) Vol. **140**, No. 1, p. 39—50 (1987).

N. MATTERN, J. HENKE, M. REIBOLD, and H. HERMANN
Study of the Structure of Annealed Amorphous $Fe_{40}Ni_{40}P_{14}B_6$
(61.40; 61.10; S1.1)
(a) Vol. **100**, No. 1, p. 37—46 (1987).

I. V. Polikarpov and V. V. Skadorov
*X-Ray Dynamical Diffraction in Crystals Being Subjected to the Action of Ultrasonic Waves
(61.10)
(b) Vol. **143**, No. 1, p. 11—17 (1987).

D. Y. Parpia, M. Safa, B. K. Tanner, and G. F. Clark
Synchrotron X-Ray Topography Observation of the Morin Transition in Highly Perfect Crystals of Haematite
(75.60; 61.10; S10)
(a) Vol. **103**, No. 1, p. 263—271 (1987).

U. Pietsch and K. Unger
*An Experimental Proof of the Valence Electron Density Variation in Silicon under High Electric Field
(61.10 and 61.50; S5.11)
(b) Vol. **143**, No. 2, p. K95—K97 (1987).

F. N. Chukhovskii, K. T. Gabrielyan, E. N. Kislovskii, and I. V. Prokopenko
*Experimental and Theoretical Study of the Elastic Bending Effect on (111) Bragg Diffraction in Silicon Crystals
(61.10 and 61.70; S5.11)
(a) Vol. **103**, No. 2, p. 381—387 (1987).

M. Wilkens, T. Ungár, and H. Mughrabi
*X-Ray Rocking-Curve Broadening of Tensile-Deformed [001]-Orientated Copper Single Crystals
(61.10 and 61.70; S1.2)
(a) Vol. **104**, No. 1, p. 157—170 (1987).

B. C. Larson and F. W. Young, Jr.
*X-Ray Diffuse Scattering Study of Irradiation Induced Dislocation Loops in Copper
(61.10 and 61.80; S1.2)
(a) Vol. **104**, No. 1, p. 273—286 (1987).

F. Aubertin, G. L. Whittle, S. J. Campbell, and U. Gonser
Hydrogen-Induced Magnetism in Zr_3Fe Hydrides
(75.30; 61.10; 76.80; S1.1)
(a) Vol. **104**, No. 1, p. 397—402 (1987).

M. Wilkens
X-Ray Line Broadening and Mean Square Strains of Straight Dislocations in Elastically Anisotropic Crystals of Cubic Symmetry
(61.70; 61.10; S1.2)
(a) Vol. **104**, No. 1, p. K1—K6 (1987).

P. A. Bezirganyan, L. G. Gasparyan, K. G. Truni, and V. P. Mkrtchyan
*X-Ray Diffraction in Interferometric Systems Consisting of Crystalline Blocks of Different Thickness
(61.10)
(a) Vol. **104**, No. 2, p. 549—567 (1987).

A. M. Afanasev, R. M. Imanov, E. Kh. Mukhamedzhanov, and A. A. Nazlukhanyan
*Coherent Compton Effect in the Grazing Bragg-Laue Geometry
(61.10; S5.11)
(a) Vol. **104**, No. 2, p. K73—K77 (1987).

61.12. Neutron determination of structures
(for specific determinations, see 61.55 to 61.80)

L. CSER, M. KOCSIS, L. KÖSZEGI, A. OLES, A. TOPOLNICKA, and J. WOLNY
Neutron Scattering Study of Carbon Fibres
(61.55; 61.12; S5)
(a) Vol. **103**, No. 1, p. 39—44 (1987).

61.14. Electron determination of structures
(for specific determinations, see 61.55 to 61.80)

R. WANG and D. LIU
*Foil Thickness Determination by CBED in the Kinematical Approximation
(61.14; S1.2; S4; S10.1)
(a) Vol. **99**, No. 1, p. 47—55 (1987).

D. LIU, R. WANG, and Y. HUANG
*Ultra-Thin Foil Thickness Determination from Large-Angle Convergent Beam Electron Diffraction Patterns
(61.14; S1.2; S4)
(a) Vol. **99**, No. 2, p. 339—344 (1987).

H. M. TAWANCY and M. O. ABOELFOTOH
Experimental Observations of Transient Phases during Long-Range Ordering to Ni_4Mo in a Ni–Mo–Fe–Cr Alloy
(64.60; 61.14; S1.2)
(a) Vol. **99**, No. 2, p. 461—466 (1987).

M. TAKEDA
*A Simplified Method for Electron Dynamical Diffraction with Crystal Symmetry Considerations
(61.14; S4; S5.11)
(a) Vol. **101**, No. 1, p. 25—36 (1987).

K. STACHULEC and A. STACHULEC
*Temperature Dependence of Low Energy Electron Diffraction
(61.14)
(b) Vol. **141**, No. 2, p. K89—K92 (1987).

G. VAN TENDELOO, D. VAN DYCK, S. KUYPERS, and S. AMELINCKX
*Electron Diffraction Effects in Mixed Layer Compounds. I. Theoretical Considerations
(61.14)
(a) Vol. **101**, No. 2, p. 339—354 (1987).

G. VAN TENDELOO, D. VAN DYCK, S. KUYPERS, H. W. ZANDBERGEN, and S. AMELINCKX
*Electron Diffraction Effects in Mixed Layer Compounds. II. Applications
(61.14; S1.2; S8.16)
(a) Vol. **102**, No. 2, p. 597—608 (1987).

V. HEERA and B. RAUSCHENBACH
*On the Calculation of the Multiplicities of Quasicrystal Diffraction Spots
(61.14)
(b) Vol. **143**, No. 1, p. K1—K4 (1987).

H. W. ZANDBERGEN, G. VAN TENDELOO, T. OKABE, and S. AMELINCKX
Electron Diffraction and Electron Microscopy of the Superconducting Compounds $Ba_2YCu_3O_{7-\delta}$
(61.60; 61.14; 61.16; S10)
(a) Vol. **103**, No. 1, p. 45—72 (1987).

H. S. KIM and S. S. SHEININ
*On the Orthogonality of Bloch Functions in the Dynamical Theory of Electron Diffraction
(61.14)
(a) Vol. **103**, No. 2, p. 349—356 (1987).

D. VAN DYCK, D. BRODDIN, J. MAHY, and S. AMELINCKX
*Electron Diffraction of Translation Interface Modulated Structures
(61.14)
(a) Vol. **103**, No. 2, p. 357—373 (1987).

H. S. KIM and S. S. SHEININ
*The Dynamical Theory of Twin Boundary Contrast for the Non-Column and Non-Symmetrical Laue Case
(61.14; 61.70)
(a) Vol. **104**, No. 2, p. 519—527 (1987).

61.16. Other determination of structures
(for specific determinations, see 61.55 to 61.80)

V. V. ARISTOV, V. I. NIKOLAICHIK, and I. I. KHODOS
*Defocus Contrast on Electron Microscope Images of Small Dislocation Loops. I. Computations on the Basis of Dynamical Theory
(61.16; 61.70)
(a) Vol. **100**, No. 1, p. 31—36 (1987).

O. SALMINEN, T. KEINONEN, R. LINDSTRÖM, and P. SILFSTEN
*The Application of Moiré Fringes to Check the Quality of Surface Relief Gratings by Using SEM
(61.16)
(a) Vol. **100**, No. 1, p. K1—K2 (1987).

R. PÉREZ
*Image Contrast Characteristics of Stacking Faults and Perfect Crystals Using the $(1\bar{1}1)$ H.C.P. Systematic Set
(61.16; 61.70; S1.2)
(a) Vol. **100**, No. 2, p. 405—414 (1987).

J. MAHY, J. VAN LANDUYT, and S. AMELINCKX
*High Resolution Electron Microscopic Analysis of the Longitudinally Deformation Modulated Structure of $TaTe_4$
(61.16; 61.60; S1.63)
(a) Vol. **102**, No. 2, p. 609—618 (1987).

R. Fabbri, G. Matteucci, and G. Pozzi
*Observations of Magnetic Domains in Thin Films by Electron Holography with a Commercial TEM with FEG
(61.16 and 75.70; S1.1)
(a) Vol. **102**, No. 2, p. K127—K129 (1987).

H. W. Zandbergen, G. van Tendeloo, T. Okabe, and S. Amelinckx
Electron Diffraction and Electron Microscopy of the Superconducting Compound $Ba_2YCu_3O_{7-\delta}$
(61.60; 61.14; 61.16; S10)
(a) Vol. **103**, No. 1, p. 45—72 (1987).

M. Ben Salem, B. Yangui, G. Schiffmacher, G. Teste de Sagey, and C. Boulesteix
*Choice of Proper Crystallographic Orientation for Good Visualization of a Glide Defect by HREM Techniques. Example: Antiphase Boundary in Monoclinic Sm_2O_3
(61.16; 61.70; S10)
(a) Vol. **103**, No. 2, p. 375—380 (1987).

S. K. Maksimov, V. V. Ovcharov, and D. I. Piskunov
*Specific Features of Electron Microscopy Images of Systems of Dilatation-Type Defects
(61.16 and 61.70)
(a) Vol. **104**, No. 2, p. 507—517 (1987).

T. Shiozawa and T. Kobayashi
High Resolution Electron Microscopic Study of Photolytic Silver in Silver Bromide
(61.80; 61.16; S1.3; S9)
(a) Vol. **104**, No. 2, p. 649—659 (1987).

61.20. Classical, semiclassical, and quantum theories of liquid structure
(for electronic states, see 71; for liquid helium, see 67)

V. Vitek and T. Egami
Atomic Level Stresses in Solids and Liquids
(61.50 and 61.70; 61.20; 61.40)
(b) Vol. **144**, No. 1, p. 145—156 (1987).

61.25. Studies of specific liquid structures

Xian-Wu Zou, Zhun-Zhi Jin, and Ya-Jun Shang
*Static Structure Factor of Non-Simple Liquid Metals Bi, Ga, Sb, and Sn
(61.25; S4)
(b) Vol. **139**, No. 2, p. 365—370 (1987).

R. V. Gopala Rao and M. Bhattacharyya
*Partial Structure of Compound-Forming Liquid Gold–Caesium Alloy. A Model Potential Approach
(61.25 and 61.55; S1.3; S2)
(b) Vol. **140**, No. 1, p. 51—61 (1987).

T. Satow, Q. Uemura, K. Matsumoto, and S. Okamura
Electronic Properties of Liquid Sb–Se and Bi–Se Alloys
(72.15; 61.25; 75.20; S4)
(b) Vol. **140**, No. 1, p. 233—242 (1987).

H. Kresse and P. Rabenstein
*Geometrical Aspects and Dipole–Dipole Interaction in a Liquid Crystalline Mixture
(61.25 and 77.20; S12)
(a) Vol. **100**, No. 1, p. K83—K85 (1987).

T. Bhattacharya and A. K. Ghosh
*Pair-Correlation Functions of Liquid Noble Metals through Born-Green-Yvonne Theory
(61.25; 65.50; S1.2; S1.3)
(b) Vol. **140**, No. 2, p. 331—334 (1987).

Xian-Wu Zou, Zhun-Zhi Jin, and Sheng-Qun Tang
*Temperature Dependence of the Static Structure Factor for Liquid Bi, Ga, Sb, and Sn
(61.25; S4)
(b) Vol. **142**, No. 1, p. 9—14 (1987).

C. Chaïb and J.-G. Gasser
The Temperature Dependence of the Resistivity of Liquid Copper
(72.15; 61.25; S1.2)
(b) Vol. **142**, No. 2, p. 531—536 (1987).

R. V. Gopala Rao and U. Bandyopadhyay
*Partial Structure Factors and Associated Correlation Functions of Molten Strontium Chloride
(61.25; S9.12)
(a) Vol. **104**, No. 1, p. 575—585 (1987).

61.30. Liquid crystals

H. P. Hinov
*DC Voltage-Controlled Pulsations of a Homotropic Nematic Layer with a Free-Slipping Liquid Crystal–Polar Liquid Boundary
(61.30 and 73.30)
(a) Vol. **101**, No. 1, p. 185—192 (1987).

S. Trimper
*Nematic Liquid Crystal in 2 + ε Dimensions
(61.30)
(b) Vol. **141**, No. 2, p. 369—377 (1987).

Zhengmin Sun, Huaqin Wang, Dirong Qiu, and Liangyu Wang
*X-Ray Diffraction Investigations in the Disc-Like Mesophase of Three Compounds of Rufigallol-hexa-n-alkanoates
(61.30; S12)
(a) Vol. **101**, No. 2, p. K93—K96 (1987).

H.-R. Trebin, L. Longa, and B. Salzgeber
*Metallic Glasses and Chiral Nematic Liquid Crystals
(61.30 and 61.40)
(b) Vol. **144**, No. 1, p. 73—84 (1987).

61.40. Amorphous and polymeric materials

O. A. Golikova
*Semiconductors with Complex Lattice and the Amorphization Problem **(Review Article)**
(61.40 and 72.20; S5)
(a) Vol. **101**, No. 2, p. 277—314 (1987).

J. Bullot and M. P. Schmidt
*Physics of Amorphous Silicon–Carbon Alloys **(Review Article)**
(61.40; 71.20; 71.55; 72.40; 73.60; 78.20; 78.55; 78.60; S6)
(b) Vol. **143**, No. 2, p. 345—418 (1987).

A. I. Olemskoi, V. E. Panin, and V. A. Petrunin
On the Transformation of Collective Modes in a Viscoelastic Medium
(62.20 and 62.50; 61.40; 66.30)
(b) Vol. **139**, No. 2, p. 387—391 (1987).

J. Mayer, K. Urban, and J. Fidler
Phase Transitions between the Quasicrystalline, Crystalline, and Amorphous Phases in Al–16 at%V
(64.70; 61.40; S4)
(a) Vol. **99**, No. 2, p. 467—473 (1987).

N. Mattern, J. Henke, M. Reibold, and H. Hermann
*Study of the Structure of Annealed Amorphous $Fe_{40}Ni_{40}P_{14}B_6$
(61.40; 61.10; S1.1)
(a) Vol. **100**, No. 1, p. 37—46 (1987).

H. Hermann, N. Mattern, and W. Matz
*Magnetic and Structural Properties of Rapidly Quenched Iron–Boron Metallic Glasses
(61.40 and 75.50; S1.1)
(b) Vol. **140**, No. 2, p. 581—588 (1987).

M. Dubiel and D. Ehrt
*Nuclear Magnetic Resonance Investigations of the Structure of Fluorophosphate Glasses
(61.40; 76.60; S11)
(a) Vol. **100**, No. 2, p. 415—422 (1987).

B. Rauschenbach and V. Heera
*Ion Irradiation Induced Reamorphization of Boron Preimplanted Iron
(61.40 and 61.80; S1.1)
(a) Vol. **100**, No. 2, p. 423—429 (1987).

A. Latgé, S. S. Makler, and E. V. Anda
*Electronic Properties of Amorphous Semiconductors. Lattice Topology Effects
(61.40 and 71.20; S5.11)
(b) Vol. **141**, No. 1, p. 163—171 (1987).

Wu Jinhua, Zhang Fangqing, and Chen Guanghua
*A New Method to Estimate the Ability of Forming Amorphous Solids
(61.40 and 64.70)
(a) Vol. **101**, No. 1, p. K1—K3 (1987).

AL. NICULA and E. CULEA
Determination of the Distribution Widths of the Spin Hamiltonian Parameters of V^{4+} Ions from Oxide Glasses
(76.30; 61.40; S10.15)
(b) Vol. **142**, No. 1, p. 265—270 (1987).

V. O. SOKOLOV and A. B. SULIMOV
*Semi-Empirical Calculation of Peroxy Linkage in Vitreous Silicon Dioxide
(61.40 and 61.70; S10.1)
(b) Vol. **142**, No. 1, p. K7—K12 (1987).

ZS. KAJCSOS
*Properties of Amorphous and Fine-Grained Alloys as Seen by Positron Annihilation
(61.40 and 78.70; S1)
(a) Vol. **102**, No. 1, p. 67—78 (1987).

E. KOLAWA, T. TYLISZCZAK, B. D. SAWICKA, J. A. SAWICKI, and M. DRWIEGA
*Studies of Local Structure and Thermal Annealing Processes in Implanted Iron–Gadolinium Alloys by Conversion Electron Mössbauer Spectroscopy
(61.40; 61.70; 76.80; S1.1; S1.4)
(a) Vol. **102**, No. 1, p. 181—192 (1987).

ZHANG BANGWEI
*A Semi-Empirical Approach to the Prediction of the Amorphous Alloys Formed by Ion Beam Mixing
(61.40)
(a) Vol. **102**, No. 1, p. 199—206 (1987).

XIAO KEQIN, DONG YUANDA, and HE YIZHEN
*A Further Study on the Crystallization of a-$Pd_{80}Si_{20}$ and an Analysis of the Structure of the Amorphous Matrix
(61.40; 64.70; S1.3)
(a) Vol. **102**, No. 1, p. 221—229 (1987).

A. V. GOLDANSKII, V. A. ONISHUK, and V. P. SHANTAROVICH
*Some Principles of the Studies of Positron Annihilation in Polymer Systems
(61.40 and 78.70; S12)
(a) Vol. **102**, No. 2, p. 559—564 (1987).

P. ARIFOV, S. VASSERMAN, and S. TISHIN
*Study of Elastomer Structure Microdefects by Positron Annihilation
(61.40 and 78.70; S12)
(a) Vol. **102**, No. 2, p. 565—570 (1987).

E. MÁRQUEZ, P. VILLARES, and R. JIMÉNEZ-GARAY
Bulk Thermal Switching in As–Se–Te Semiconducting Glasses with a Coplanar Point Electrode Configuration
(72.80; 61.40; S8.16)
(a) Vol. **102**, No. 2, p. 741—749 (1987).

ZHIWEN XIE, DIHUA DING, and DECHENG TIAN
Magnetic After-Effects in $(Fe_{1-x}Ga_x)_{84}B_{16}$ Amorphous Alloys
(75.50; 61.40; S1.1)
(a) Vol. **102**, No. 2, p. 781—789 (1987).

61.50. Crystalline state (including molecular motions in solids)

(for magnetic structure and spin systems, see 75.25)

J. Shanker and M. Kumar
*Ion-Dependent and Crystal-Independent Interionic Potentials **(Review Article)**
(61.50; 61.70; 62.20; S9.11)
(b) Vol. **142**, No. 2, p. 325—352 (1987).

A. G. Pelmenshchikov, G. M. Zhidomirov, S. Beran, and J. Tiňo
*Molecular Approach in the Study of the Processes of Formation of the Lattice of Oxides. Characteristic Features of Coordination of AlO_3 Structural Elements
(61.50; S10)
(a) Vol. **99**, No. 1, p. 57—64 (1987).

A. I. Ustinov, A. Yu. Gaevskii, A. D. Rood, V. S. Skorodzievskii, and K. V. Chuistov
*The Influence of the Electron Subsystem and Kinetic Factors on the Formation of One-Dimensionally Disordered Structures with Long-Range Correlation
(61.50; S1.2)
(a) Vol. **99**, No. 1, p. 81—90 (1987).

A. M. Balagurov, A. I. Beskrovnii, B. N. Savenko, B. V. Merinov, M. Dlouhá, S. Vratislav, and Z. Jirák
*The Room Temperature Structure of Deuterated $CsHSO_4$ and $CsHSeO_4$
(61.50; S11.1)
(a) Vol. **100**, No. 1, p. K3—K7 (1987).

A. I. Gusev and A. A. Rempel
*Calculating the Energy Parameters for the CV and OPF Methods
(61.50; 64.60; S1.61)
(b) Vol. **140**, No. 2, p. 335—346 (1987).

L. Firlej, A. Graja, A. Rajchel, K. Woźniak, and T. M. Krygowski
*The 7,7′, 8,8′-Tetracyano-p-Quinodimethanide Salt of the N-n Propyl Quinoxalinium Cation — Its Structure, Conductivity, and Triplet Spin Excitons
(61.50 and 72.80; 76.30; S12.1)
(b) Vol. **140**, No. 2, p. 437—447 (1987).

V. Kumar
*Bond Susceptibility and Ionicity in $A^{III}B^{V}C_4^{VI}$ Compounds
(61.50 and 77.20; S10; S10.1)
(b) Vol. **141**, No. 2, p. K157—K162 (1987).

E. Carrillo H., A. Carrillo H., and J. Rubio O.
*A Simple Geometric Model for Predicting Coherence between Matrix and Precipitated Phases in Alkali Halides Doped with Divalent Impurities
(61.50; 61.70; S9.11)
(a) Vol. **101**, No. 2, p. 315—327 (1987).

S. Abdi, Y. Mlik, and M. H. Ben Ghozlen
*Group-Theoretical Analysis of the Symmetry Changes in R_2MX_4 Compounds
(61.50; 64.70; S9.16)
(b) Vol. **142**, No. 1, p. 27—34 (1987).

V. VITEK and T. EGAMI
*Atomic Level Stresses in Solids and Liquids
(61.50 and 61.70; 61.20; 61.40)
(b) Vol. **144**, No. 1, p. 145—156 (1987).

F. ERNST and P. HAASEN
*The Decomposition Kinetics of Al–1 at% Ag at 413 K Studied by HREM
(61.50 and 64.75; S4)
(a) Vol. **104**, No. 1, p. 403—416 (1987).

N. NAFARI and A. M. AWIN
*The Calculation of the Madelung Constant for NaCl Structure Using the Method of Images
(61.50)
(a) Vol. **104**, No. 1, p. K21—K25 (1987).

P. MOHAZZABI and F. BEHROOZI
*A Reexamination of the Continuum Approach to the Calculation of Lattice Sums
(61.50)
(b) Vol. **144**, No. 2, p. 459—462 (1987).

M. KUMAR, D. HANS, A. K. SHARMA, and J. SHANKER
Structural Phase Transition of Alkali Halides Using a Modified Interaction Model
(64.60; 61.50; S9.11)
(b) Vol. **144**, No. 2, p. 543—548 (1987).

K. S. ALEKSANDROV, B. V. BEZNOSIKOV, and S. V. MISYUL
*Successive Phase Transitions in Crystals of K_2MgF_4-Type Structure
(61.50; 64.60; S9.16)
(a) Vol. **104**, No. 2, p. 529—543 (1987).

I. G. MARGVELASHVILI and Z. K. SARALIDZE
*Specific Features of the Point Defect Equilibrium Distribution and Kinetics in the Ordered Two-Component (Non-Ionic) Crystals
(61.50 and 61.70)
(a) Vol. **104**, No. 2, p. 587—596 (1987).

61.55. Specific structures of elements and alloys

A. OSTRASZ and M. SZUSKIEWICZ
3d-Like Electron Localization and Charge Transfer in Hydrogenated V–Fe Alloys
(71.25; 61.55; 76.80; S1.1; S1.2)
(b) Vol. **139**, No. 1, p. 205—211 (1987).

H. LÖFFLER, G. WENDROCK, and K. PAWELZYK
*The Influence of Small Mg Additions on the Stability of Guinier-Preston Zones in Al–4.5 at% Zn–x Mg Alloys
(61.55; 61.10; 64.70; S4)
(a) Vol. **99**, No. 1, p. 73—80 (1987).

S. P. CHENAKIN, V. T. CHEREPIN, A. L. PIVOVAROV, and M. A. VASILEV
*Temperature Dependence of Secondary Ion Emission from Iron-Based Crystalline and Amorphous Alloys
(61.55 and 79.20; S1.1)
(a) Vol. **102**, No. 1, p. 231—240 (1987).

J. MÄKI
*On the Effect of Substitutional Elements on the Out-Of-Step Distance in the Long Period Ordered Ag_3Mg-Based Alloys
(61.55; S1.3)
(a) Vol. **102**, No. 2, p. 589—596 (1987).

T. TAKASUGI and O. IZUMI
Defect Structures in Co-Rich CoTi Intermetallic Compound
(61.70; 61.55; S1.2)
(a) Vol. **102**, No. 2, p. 697—702 (1987).

H.-G. FRITSCHE
*The Change of Icosahedron–Cuboctahedron Structures of Growing Argon Crystallites
(61.55; S5.2)
(b) Vol. **143**, No. 1, p. K11—K14 (1987).

R. SCHÜLBE and U. SCHMIDT
*Compilation of the Results of Reversion Investigations of Al–1.87 at% Cu Alloys
(61.55; 62.20; S4)
(a) Vol. **103**, No. 1, p. 29—38 (1987).

L. CSER, M. KOCSIS, L. KÖSZEGI, A. OLES, A. TOPOLNICKA, and J. WOLNY
*Neutron Scattering Study of Carbon Fibres
(61.55; 61.12; S5)
(a) Vol. **103**, No. 1, p. 39—44 (1987).

R. V. G. RAO and M. BHATTACHARYYA
*Caesium–Gold as an Ionic Alloy and Its Structural Properties
(61.55; S1.3; S2)
(a) Vol. **103**, No. 1, p. 79—84 (1987).

I. DÉZSI, H. ENGELMANN, U. GONSER, and G. LANGOUCHE
*Cobalt–Germanium Structures Investigated by Mössbauer Spectroscopy
(61.55; 61.80; 76.80; S1.61; S5.12)
(a) Vol. **104**, No. 1, p. 417—426 (1987).

P. KRATOCHVÍL, M. MANDULA, J. MENCL, J. PEŠIČKA, and B. SMOLA
*The Constitution of the Cu–9 Ni–6 Sn Alloy Determined by Changes of the Strength and by Structure Determination during Annealing
(61.55; 62.20; S1.2)
(a) Vol. **104**, No. 2, p. 597—602 (1987).

61.60. Specific structures of inorganic compounds

V. I. Voronkova and V. K. Yanovskii
*The Structure and Ferroelectric Properties of Layered Compounds in the System $Bi_4Ti_3O_{12}$–Bi_2WO_6
(61.60; 64.70; 77.80; S10.15)
(a) Vol. **101**, No. 1, p. 45—50 (1987).

J. S. Diez Viñuela and C. Otero Areán
*Distribution of Copper Ions among Octahedral and Tetrahedral Sites in $Cu_xMg_{1-x}Al_2O_4$ Spinels
(61.60; S11.2)
(a) Vol. **101**, No. 1, p. 57—61 (1987).

S. N. Toubektsis, J. A. Kalomiros, L. V. Giobliakis, K. M. Paraskevopoulos, A. N. Anagnostopoulos, and E. K. Polychroniadis
*On the Growth and Characterization of the System $TlBiSe_2$–$TlBiTe_2$
(61.60; 72.80; 78.20; S8.16)
(a) Vol. **101**, No. 2, p. 355—360 (1987).

F. W. Boswell, A. Prodan, J. C. Bennett, J. M. Corbett, and L. G. Hiltz
*A Phase Transition and Domain Boundaries in the Modulation Structure of $TaTe_4$
(61.60; 61.70;, S8)
(a) Vol. **102**, No. 1, p. 207—220 (1987).

M. Piotrowski, L. Chełmicki, B. Paluchowska, K. Blinowski, and M. Konwicki
Magnetic Structure of the Li–Ti–Zn Spinel Ferrites by Polarized Neutrons
(75.50; 61.60; S11.2)
(a) Vol. **102**, No. 1, p. 375—380 (1987).

J. Mahy, J. van Landuyt, and S. Amelinckx
High Resolution Electron Microscopic Analysis of the Longitudinally Deformation Modulated Structure of $TaTe_4$
(61.16; 61.60; S1.63)
(a) Vol. **102**, No. 2, p. 609—618 (1987).

C. Jäger, S. Barth, A. Feltz, and G. Scheler
Studies of the Structure of Mixed Crystals of the System $Na_{1+2x}Mg_xZr_{2-x}(PO_4)_3$ by ^{31}P-MAS-NMR
(76.60; 61.60; S11.1)
(a) Vol. **102**, No. 2, p. 791—802 (1987).

Z. Jirak, E. Pollert, A. Tříska, and S. Vratislav
*Structure and the Orthorhombic–Tetragonal Transition in $YBa_2Cu_3O_7$
(61.60 and 64.70; S10)
(a) Vol. **102**, No. 2, p. K61—K66 (1987).

S. Manhas, K. C. Khulbe, D. J. S. Beckett, G. Lamarche, and J. C. Woolley
Lattice Parameters, Energy Gaps, and Magnetic Properties of the $Cd_xHg_yMn_zTe$ Alloy System
(75.30; 61.60; 71.25; S8.16)
(b) Vol. **143**, No. 1, p. 267—274 (1987).

H. W. Zandbergen, G. van Tendeloo, T. Okabe, and S. Amelinckx
*Electron Diffraction and Electron Microscopy of the Superconducting Compounds $Ba_2YCu_3O_{7-\delta}$
(61.60; 61.14; 61.16; S10)
(a) Vol. **103**, No. 1, p. 45—72 (1987).

H. Ketata, M. H. Ben Ghozlen, A. Daoud, and I. Pabst
*Low-Temperature Structure ($T = 235$ K) and Structural Phase Transition in $[(C_2H_5)_4N]_2SnCl_6$
(61.60; 64.70; S9.15)
(a) Vol. **103**, No. 1, p. 73—78 (1987).

A. M. Balagurov, F. Prokert, and B. N. Savenko
Phase Transition Effects in the Incommensurate Modulated Single Crystal of $Sr_{0.7}Ba_{0.3}Nb_2O_6$
(64.70; 61.60; S11.1)
(a) Vol. **103**, No. 1, p. 131—144 (1987).

A. A. Kaminskii, B. V. Mill, A. V. Butashin, E. L. Belokoneva, and K. Kurbanov
Germanates with $NdAlGe_2O_7$-Type Structure. Synthesis, Crystal Structure, Absorption–Luminescence Properties, and Stimulated Emission of Their Activator, Nd^{3+} Ions
(78.30 and 78.55; 61.60; 78.45; S10)
(a) Vol. **103**, No. 2, p. 575—592 (1987).

O. Uemura, Y. Takashima, K. Muraishi, and T. Satow
*A Change in Local Order of Atmorphous $CdGeAs_2$ on the Glass–Crystal Transition and the High Temperature Modification of Crystalline Phase
(61.50 and 61.60; S7.16)
(a) Vol. **103**, No. 2, p. K69—K74 (1987).

K. S. Aleksandrov, S. V. Melnikova, and S. V. Misyul
*Successive Phase Transitions in Ferroelastic Cs_2RbDyF_6 with Elpasolite-Type Structure
(61.60, and 64.70; S9.16)
(a) Vol. **104**, No. 2, p. 545—548 (1987).

E. Wolska and A. Wolniewicz
*On the Aluminium-Substituted Magnetites
(61.60 and 64.75; S10.15)
(a) Vol. **104**, No. 2, p. 569—574 (1987).

K. D. Bronsema and J. Mahy
*One-, Two-, and Three-Dimensional Order in and Four-Dimensional Description of $K_xV_5S_8$ ($0.5 \leqq x \leqq 0.7$)
(61.60; S8.16)
(a) Vol. **104**, No. 2, p. 603—618 (1987).

61.65 Specific structures of organic compounds

J. Mertsching
*Peierls Phases in Polyacene
(61.65 and 71.25; S12.1)
(b) Vol. **143**, No. 1, p. K5—K10 (1987).

61.70. Defects in crystals

(see also 61.80. Radiation damage 62. Mechanical and acoustical properties, 71.55. Impurities and defect levels, 76.30. EPR of color centers and other defects, 78.50. Impurity and defect absorption in solids)

J. SHANKER and M. KUMAR
Ion-Dependent and Crystal-Independent Interionic Potentials **(Review Article)**
(61.50; 61.70; 62.20; S9.11)
(b) Vol. **142**, No. 2, p. 325—352 (1987).

G. DLUBEK and R. KRAUSE
*Positron Studies of Defects in III–V Semiconductor Compounds **(Review Article)**
(61.70 and 78.70; S7.1)
(a) Vol. **102**, No. 2, p. 443—479 (1987).

I. A. OVIDKO and A. E. ROMANOV
*Topological Excitations (Defects, Solitons, Textures, Frustrations) in condensed Media **(Review Article)**
(61.70; 64.60; 71.45; 75.30)
(a) Vol. **104**, No. 1, p. 13—45 (1987).

H.-J. FÖCKEL
*Stationary and Resonance States at the Core of a Screw Dislocation and Their Contributions to Electrical Resistivity
(61.70 and 71.55; 72.15; S2)
(b) Vol. **139**, No. 1, p. 59—66 (1987).

A. GHORAI and S. K. SEN
*Calculations on Relaxations Due to a Vacancy, Binding Energy of a Vacancy–Vacancy Pair, and Formation Energy of a Vacancy in Thorium
(61.70; 63.20; S1.5)
(b) Vol. **139**, No. 1, p. 77—83 (1987).

S. LIU, L. Y. XIONG, and C. W. LUNG
*The Image Force Modified Dislocation Distribution in a Cracked Finite Width Material
(61.70)
(a) Vol. **99**, No. 1, p. 97—104 (1987).

B. KANTHA REDDY
*Annealing and Ageing Studies in Quenched $KBr:Ba^{2+}$ Single Crystals
(61.70; 62.20; S9.11)
(a) Vol. **99**, No. 1, p. K7—K10 (1987).

W. SEIFERT and M. KITTLER
*Negative (Bright) EBIC Contrast at Oxygen-Induced Defects in Silicon
(61.70; 72.20; S5.11)
(a) Vol. **99**, No. 1, p. K11—K14 (1987).

J. TATARKIEWICZ and K. WINER
*Vibrational Properties of Hydrogen in Proton-Implanted Silicon
(61.70 and 63.20; 78.30; S5.11)
(b) Vol. **139**, No. 2, p. 409—415 (1987).

V. Holý and K. T. Gabrielyan
Dyson and Bethe-Salpeter Equations for Dynamical X-Ray Diffraction in Crystals with Randomly Placed Defects
(61.10; 61.70)
(b) Vol. **140**, No. 1, p. 39—50 (1987).

M. M. Merkin and A. A. Tsertsvadze
Exoemission Theory: The Exciton Model
(79.75; 61.70; 71.35; S9.11)
(b) Vol. **140**, No. 1, p. 73—82 (1987).

P. H. E. Meijer, P. Pecheur, and G. Toussaint
Electronic Structure of the Cd Vaçancy in CdTe
(71.55; 61.70; 71.25; S8.13)
(b) Vol. **140**, No. 1, p. 155—162 (1987).

V. V. Aristov, V. I. Nikolaichik, and I. I. Khodos
Defocus Contrast on Electron Microscope Images of Small Dislocation Loops. I. Computations on the Basis of Dynamical Theory
(61.16; 61.70)
(a) Vol. **100**, No. 1, p. 31—36 (1987).

G. Vanderschaeve
*Cross-Slip of Partial Dislocations via the Stair Rod Mode. The Zigzag Propagation of Deformation Microtwins in Ordering Alloys
(61.70; S1.2)
(a) Vol. **100**, No. 1, p. 59—68 (1987).

K. Schmalz, F.-G. Kirscht, S. Niese, H. Richter, M. Kittler, W. Seiffert, I. Babanskaya, H. Klose, K. Tittelbach-Helmrich, and J. Schöneich
*On the Intrinsic Gettering in Fe-Contaminated Cz–Si
(61.70; 71.55; S5.11)
(a) Vol. **100**, No. 1, p. 69—85 (1987).

W. Th. M. Buters, A. van Veen, and A. van den Beukel
*Some Results on Helium Trapping in Undeformed and Cold Worked Single Crystalline Copper. A Comparison with Molybdenum and Nickel
(61.70; 68.45; 79.20; S1.2)
(a) Vol. **100**, No. 1, p. 87—94 (1987).

P. Zaumseil, U. Winter, F. Cembali, M. Servidori, and Z. Sourek
*Determination of Dislocation Loop Size and Density in Ion Implanted and Annealed Silicon by Simulation of Triple Crystal X-Ray Rocking Curves
(61.70; 61.10; S5.11)
(a) Vol. **100**, No. 1, p. 95—104 (1987).

M. T. Sprackling and M. E. Bahrololoom
*The Role of Iodide in the Strain-Ageing of Silver Iodobromide
(61.70; 62.20; S9.15)
(a) Vol. **100**, No. 1, p. 105—108 (1987).

M. Gabbay, A. Vincent, and G. Fantozzi
Amplitude-Dependent Internal Friction Due to Thermomechanical Breakaway of Dislocations from Pinning Point Defects
(62.40; 61.70)
(a) Vol. **100**, No. 1, p. 121—137 (1987).

K. Iwasaki
*Strain Amplitude Dependence of the Grain Boundary Internal Friction Peak of Dilute Aluminium Alloys
(61.70; 62.40; S4)
(a) Vol. **100**, No. 2, p. 453—459 (1987).

K. Schmalz, F.-G. Kirscht, H. Klose, H. Richter, and K. Tittelbach-Helmrich
DLTS Study on Deep Level Defects in Cz-p-Si Due to Heat Treatment at 600 to 900 °C
(71.55; 61.70; S5.11)
(a) Vol. **100**, No. 2, p. 567—582 (1987).

K. Schmalz, R. Krause, H. Richter, and K. Tittelbach-Helmrich
*On the Hydrogen Passivation of Thermal Defects in Czochralski Silicon
(61.70; S5.11)
(a) Vol. **100**, No. 2, p. K123—K128 (1987).

A. Mokrani, G. Vacquier et C. Boulesteix
*Défants de structure d'un matériau en feuillet: $NbSe_2$
(Structural Defects in Layer Material: $NbSe_2$)
(61.70; S163)
(a) Vol. **100**, No. 2, p. K129—K134 (1987).
Erratum
(a) Vol. **104**, No. 2, p. K147 (1987).

V. I. Barbashov, V. I. Zaitsev, V. M. Mostovoi, and V. I. Alekseenko
*Mobility of 60° Dislocations in Indium Antimonide on Reversing the Load
(61.70; S7.13)
(a) Vol. **100**, No. 2, p. K135—K138 (1987).

A. Zaiour, M. Hage-Ali, J. M. Koebel, A. Bentz, and P. Siffert
*Determination of Segregation Coefficients of a Series of Impurities in Tellurium
(61.70 and 64.75; S5)
(a) Vol. **100**, No. 2, p. K139—K143 (1987).

B. G. Dick
*Microscopic Models for Substitutional OH^- and CN^- Defects in Alkali-Halide Crystals
(61.70; S9.11)
(b) Vol. **141**, No. 1, p. 61—70 (1987).

M. Kohyama
*Structures and Energies of Symmetrical ⟨001⟩ Tilt Grain Boundaries in Silicon
(61.70; S5.11)
(b) Vol. **141**, No. 1, p. 71—83 (1987).

S. V. Gorbunov, A. V. Kruzhalov, and M. J. Springis
Optical Properties of the F-Centre in Beryllium Oxide
(78.50 and 78.55; 61.70; 71.55; S10.1)
(b) Vol. **141**, No. 1, p. 293—301 (1987).

S. V. Bulyarskii and V. P. Oleinikov
*Thermodynamical Evaluation of Point Defect Density and Impurity Solubility in Compound Semiconductors
(61.70)
(b) Vol. **141**, No. 1, p. K7—K10 (1987).

M. Reiche and O. Breitenstein
*The Role of Silicon Self-Interstitials in the Formation of Thermally Induced Rod-Like Defects in Cz Silicon
(61.70; S5.11)
(a) Vol. **101**, No. 2, p. K97—K99 (1987).

L. I. Murin
*On the Electrical Activity of the C_i–O_i Complex in Silicon
(61.70 and 71.55; 61.80; S5.11)
(a) Vol. **101**, No. 2, p. K107—K110 (1987).

H. Cebulla, E. Erben, T. Gessner, and E. Vetter
*Induced Tantalum Silicide Formation by Ar^+ Ion Implantation
(61.70 and 68.55; S1.61)
(a) Vol. **101**, No. 2, p. K125—K127 (1987).

G. G. Bokuchava and A. I. Tugushi
Ultrasonic Absorption in Hydrogen Containing Metals
(62.80; 61.70)
(b) Vol. **142**, No. 1, p. K1—K6 (1987).

V. O. Sokolov and A. B. Sulimov
*Semi-Empirical Calculation of Peroxy Linkage in Vitreous Silicon Dioxide
(61.40 and 61.70; S10.1)
(b) Vol. **142**, No. 1, p. K7—K12 (1987).

K. Matsuura, S. Kishida, I. Tsurumi, and M. Kitagawa
Electron-Irradiation Effects on the F^+ Bands and the Near-Infrared Emissions in ZnS Crystals
(78.50 and 78.55; 61.70; S8.11)
(b) Vol. **142**, No. 1, p. K79—K83 (1987).

H.-E. Schaefer
*Investigation of Thermal Equilibrium Vacancies in Metals by Positron Annihilation
(61.70; 71.60; 78.70; S1; S4)
(a) Vol. **102**, No. 1, p. 47—65 (1987).

C. Hidalgo, S. Linderoth, and N. de Diego
Positron–Dislocation Interaction in Deformed Zn and Cd
(71.60; 61.70; S1; S1.2)
(a) Vol. **102**, No. 1, p. 113—117 (1987).

R. Würschum, M. Scheytt, and H.-E. Schaefer
Nanocrystalline Metals and Semiconductors Studied by Positron Lifetime Spectroscopy
(61.55 and 78.70; 61.70; S1.1; S1.2; S1.3; S5.11)
(a) Vol. **102**, No. 1, p. 119—126 (1987).

A. Belaidi, H. P. Leighly, P. G. Coleman, and R. N. West
A Positron Study of Defect Formation in Cd–Mg Alloys
(64.75 and 78.70; 61.70; S1)
(a) Vol. **102**, No. 1, p. 127—132 (1987).

A. M. Mazzone
*Monte Carlo Simulation of Channeled and Random Profiles of Heavy Ions Implanted in Silicon at High Energy (1.2 MeV)
(61.70; 61.80; S5.11)
(a) Vol. **102**, No. 1, p. K13—K18 (1987).

M. L. Mukherjee and R. Capelletti
*X-Ray Induced Relaxation Band Due to Higher Aggregates in KCl:Pb
(61.70; 77.30; 78.50; S9.11)
(b) Vol. **142**, No. 2, p. 361—366 (1987).

G. Elbers, S. Remme, and G. Lehmann
*EPR and Optical Absorption of Cr^{3+} in CsCl and CsBr
(61.70; 76.30; 78.50; S9.11)
(b) Vol. **142**, No. 2, p. 367—377 (1987).

K. Matsuura, S. Kishida, K. Yoshida, and I. Tsurumi
*Optical Bleaching of the F^+ Optical Absorption Bands in ZnS Crystals
(61.70 and 78.50; S8.11)
(b) Vol. **142**, No. 2, p. 617—627 (1987).

J. Tatarkiewicz, A. M. Witowski, and E. Jahne
*Hydrogen Vibrations in CdTe
(61.70 and 63.20; 78.30; S8.13)
(b) Vol. **142**, No. 2, p. K101—K104 (1987).

A. F. Khokhlov, A. I. Mashin, A. V. Ershov, Yu. A. Mordvinova, and D. A. Pavlov
The Properties of Amorphous Silicon Doped with Isovalent Impurities
(73.60 and 78.65; 61.70; S5.11)
(b) Vol. **142**, No. 2, p. K125—K129 (1987).

S. Dannefaer
*Defects and Oxygen in Silicon Studied by Positrons
(61.70 and 78.70; S5.11)
(a) Vol. **102**, No. 2, p. 481—491 (1987).

T. Figielski
*Non-Stoichiometric Defects in GaAs Crystals
(61.70; S7.12)
(a) Vol. **102**, No. 2, p. 493—498 (1987).

J. C. Bourgoin, H. J. von Bardeleben, and D. Stievenard
Irradiation Induced Defects in III–V Semiconductor Compounds
(61.80; 61.70; 78.70; S7.1)
(a) Vol. **102**, No. 2, p. 499—510 (1987).

T. Goworek
*Defects in Molecular Solids. Positron Studies
(61.70 and 78.70; S12)
(a) Vol. **102**, No. 2, p. 511—526 (1987).

G. Dlubek, R. Krause, M. Mühlberg, and M. Schenk
*Positron Study of Nonstoichiometric Vacancies in as-Grown p-PbTe Single Crystals
(61.70 and 78.70; S8)
(a) Vol. **102**, No. 2, p. K67—K70 (1987).

P. K. Khabibullaev and S. Dzhumanov
*The Coexistence of Unstable, Metastable, and Separated Frenkel Pair Defects in Solids. II. Theory of the Impact Mechanism for Defect Formation in Metals
(61.70; S1.2; S1.3)
(b) Vol. **143**, No. 1, p. 19—35 (1987).

J. Poźniak, G. Berg, and T. Galbatar
*Aggregation Kinetics of Divalent Impurity–Vacancy Dipoles in Alkali-Halide Matrix. I. Theoretical Approach
(61.70)
(b) Vol. **143**, No. 1, p. 37—42 (1987).

A. Richter, A. E. Romanov, W. Pompe, and V. I. Vladimirov
*On the Screening Length of Disclinations in Amorphous Structures
(61.70; 61.40)
(b) Vol. **143**, No. 1, p. 43—53 (1987).

S. A. Gundersen and J. Lothe
A New Method for Numerical Calculations in Anisotropic Elasticity Problems
(62.30; 61.70)
(b) Vol. **143**, No. 1, p. 73—85 (1987).

S. L. Drechsler, E. Heiner, and J. Malek
Selfconsistent Discrete Equations for Nonlinear Defects in Peierls Systems with Broken Electron–Hole Symmetry
(71.50; 61.70; 71.45; S12.1)
(b) Vol. **143**, No. 1, p. 185—193 (1987).

K. Matsuura, S. Kishida, Y. Fukata, and I. Tsurumi
Correlation between the 1.1 and 1.45 eV Emission Bands and the F^+ Optical Absorption Bands in ZnS Crystals
(78.50 and 78.55; 61.70; S8.11)
(b) Vol. **143**, No. 1, p. 275—280 (1987).

Y. Nakada and T. Imura
*Thermal Stress and Stacking Fault Energy Estimated by TEM Observations of Grown-in Dislocations in Si-Doped GaAs Single Crystals
(61.70; S7.12)
(a) Vol. **103**, No. 1, p. 85—91 (1987).

H. Opyrchał and M. Manfredi
*Aggregation Processes in NaCl:Eu^{2+} System. Eu^{2+} Concentration Dependence of the Characteristics of the Aggregates Formed at Room Temperature
(61.70; 78.55; S9.11)
(a) Vol. **103**, No. 1, p. 93—99 (1987).

R. Würschum and H.-E. Schaefer
*Defects in Low-Temperature Electron-Irradiated GaAs Studied by Positrons
(61.70; 61.80; 78.70; S7.12)
(a) Vol. **103**, No. 1, p. 101—105 (1987).

N. Shukla and G. C. Trigunayat
Study of Structural Defects in Melt-Grown Cadmium Iodide Crystals
(61.50; 61.70; S9)
(a) Vol. **103**, No. 2, p. 337—347 (1987).

M. Ben Salem, B. Yangui, G. Schiffmacher, G. Teste de Sagey, and C. Boulesteix
Choice of Proper Crystallographic Orientation for Good Visualization of a Glide Defect by HREM Techniques. Example: Antiphase Boundary in Monoclinic Sm_2O_3
(61.16; 61.70; S10)
(a) Vol. **103**, No. 2, p. 375—380 (1987).

F. N. Chukhovskii, K. T. Gabrielyan, E. N. Kislovskii, and I. V. Prokopenko
*Experimental and Theoretical Study of the Elastic Bending Effect on (111) Bragg Diffraction in Silicon Crystals
(61.10 and 61.70; S5.11)
(a) Vol. **103**, No. 2, p. 381—387 (1987).

Y. B. Xia and Z. G. Wang
*Evolution of Dislocation Structures during Low Cycle Fatigue in a Cr–Mn–N Stainless Steel
(61.70 and 62.20; S1.1)
(a) Vol. **103**, No. 2, p. 389—395 (1987).

Cs. Szeles, K. Süvegh, Z. Homonnay, and A. Vértes
*Positron Lifetime and Mössbauer Spectroscopy Study of Vacancy–Tin Interaction in Dilute Al–Sn Alloys
(61.70; 76.80; 78.70; S4)
(a) Vol. **103**, No. 2, p. 397—401 (1987).

I. Kanazawa, H. Murakami, and M. Doyama
*The Migration Stage of Divacancies in Quenched Pure Aluminium, Dilute Aluminium Alloys, and Pure Copper Studied by Positron Annihilation
(61.70; 78.70; S1.2; S4)
(a) Vol. **103**, No. 2, p. 403—411 (1987).

J. Reichel and S. Sevcik
*Neutralization of Acceptors and Formation of Agglomerates in Silicon Wafers Due to Intrinsic Point Defects Created by Chemomechanical Polishing and by Quenching
(61.70; S5.11)
(a) Vol. **103**, No. 2, p. 413—420 (1987).

Y. Nakada and T. Imura
Influence of Dopants and Deformation Temperature on the Strength of GaAs
(62.20; 61.70; S7.12)
(a) Vol. **103**, No. 2, p. 435—442 (1987).

S. Ohtaki
The Nature of Stage II_A Recovery in Cold-Worked Aluminium
(62.20; 61.70; 72.15; S4)
(a) Vol. **103**, No. 2, p. 449—457 (1987).

S. Kishida, K. Matsuura, H. Nagase, and I. Tsurumi
Annealing Characteristics and Temperature Dependence of the Photosensitive 1.9 eV Absorption Band in Zn-Treated ZnSe Crystals
(78.50; 61.70; S8.12)
(a) Vol. **103**, No. 2, p. 613—618 (1987).

G. Vogl
*On the Motion of Solute Atoms in Cages and Loops in Metal Lattices Studied with Nuclear Methods
(61.70; 66.30; 76.80; S1; S1.3; S4)
(b) Vol. **144**, No. 1, p. 259—269 (1987).

N. Smetniansky-De Grande, E. J. Savino, and C. N. Tomé
*Stress Induced Anisotropic Diffusion of Intrinsic Point Defects towards Dislocations in H.C.P. Crystals
(61.70 and 66.30)
(b) Vol. **144**, No. 1, p. 271—286 (1987).

P. Vargas, H. Kronmüller, and M. C. Böhm
*H_2 Dissociation in Nickel. A Cluster Model
(61.70; 66.30; S1.2)
(b) Vol. **144**, No. 1, p. 305—313 (1987).

G. Neumann
*A Model for the Calculation of Monovacancy and Divacancy Contributions to the Impurity Diffusion in Noble Metals
(61.70 and 66.30; S1.2; S1.3)
(b) Vol. **144**, No. 1, p. 329—341 (1987).

F. R. N. Nabarro
Plastic Deformation in Tension and in Torsion
(62.20; 61.70)
(a) Vol. **104**, No. 1, p. 47—50 (1987).

L. M. Clarebrough and C. T. Forwood
*Formation of Stacking-Fault Tetrahedra by Grain Boundary Movement
(61.70; S1.2)
(a) Vol. **104**, No. 1, p. 51—62 (1987).

M. Werner
Temperature and Strain-Rate Dependence of the Flow Stress of Ultrapure Tantalum Single Crystals
(62.20; 61.70; S1)
(a) Vol. **104**, No. 1, p. 63—78 (1987).

B. Šesták, Z. Vicherková, V. Novák, S. Libovický, and J. Brádler
Cyclic Plastic Deformation of Fe–13.6 wt% Cr Single Crystals
(62.20; 61.70; S1.1)
(a) Vol. **104**, No. 1, p. 79—93 (1987).

K. Differt, U. Essmann, and H. Mughrabi
Models of Particle Destruction in Fatigued Precipitation-Hardened Alloys
(62.20; 61.70; S1.2)
(a) Vol. **104**, No. 1, p. 95—106 (1987).

H. Mughrabi
*The Long-Range Internal Stress Field in the Dislocation Wall Structure of Persistent Slip Bands
(61.70 and 62.20; S1.2)
(a) Vol. **104**, No. 1, p. 107—120 (1987).

H.-E. Schaefer and F. Banhart
*Thermal Equilibrium Vacancies in Platinum Studied by Positron Annihilation
(61.70; 71.60; 78.70; S1.3)
(a) Vol. **104**, No. 1, p. 263—272 (1987).

M.-P. Macht, A. Müller, V. Naundorf, and H. Wollenberger
*Determination of Temperature-Dependent Sink Concentrations in Ion-Irradiated Nickel and Copper by Diffusion Measurements
(61.70; 61.80; 66.30; S1.2)
(a) Vol. **104**, No. 1, p. 287—295 (1987).

A. Bartels, D. Bartusel, and K. Lücke
*On the Nature of Frenkel Defects in Concentrated AuAg Alloys
(61.70; 72.12; S1.3)
(a) Vol. **104**, No. 1, p. 315—328 (1987).

F. Phillipp
*Growth Characteristics of Dislocation Loops in Molybdenum Studied by In-Situ High-Voltage Electron Microscopy
(61.70; S1)
(a) Vol. **104**, No. 1, p. 329—342 (1987).

F. Walz and H. J. Blythe
*Binding Enthalpies of Carbon–Vacancy Complexes in α-Fe and Their Compatibility with Different Annealing Models
(61.70; 61.80; 71.55; 75.60; S1.1)
(a) Vol. **104**, No. 1, p. 343—356 (1987).

W. Benoit, M. Bujard, and G. Gremaud
*Kink Dynamics in F.C.C. Metals
(61.70; 62.20; 62.40; 62.80; S4)
(a) Vol. **104**, No. 1, p. 427—441 (1987).

H. Indrawirawan, O. Buck, and O. N. Carlson
Substitutional–Interstitial Solute Interactions in Niobium–Vanadium–Oxygen Alloys
(62.40; 61.70; 66.30; S1; S1.2)
(a) Vol. **104**, No. 1, p. 443—451 (1987).

N. Kuramochi, H. Mizubayashi, and S. Okuda
*Low-Temperature Dislocation Internal Friction in Hydrogen-Free Niobium
(61.70; 62.40; S1)
(a) Vol. **104**, No. 1, p. 453—467 (1987).

M. Wilkens
*X-Ray Line Broadening and Mean Square Strains of Straight Dislocations in Elastically Anisotropic Crystals of Cubic Symmetry
(61.70; 61.10; S1.2)
(a) Vol. **104**, No. 1, p. K1—K6 (1987).

E. RADZHABOV
Vacuum Ultraviolet Absorption and Dissociation of H_a^- and OH^- Centers in NaF
(78.50; 61.70; 71.55; S9.11)
(b) Vol. **144**, No. 2, p. K151—K155 (1987).

S. K. MAKSIMOV, V. V. OVCHAROV, and D. I. PISKUNOV
*Specific Features of Electron Microscopy Images of Systems of Dilatation-Type Defects
(61.16 and 61.70)
(a) Vol. **104**, No. 2, p. 507—517 (1987).

H. S. KIM and S. S. SHEININ
The Dynamical Theory of Twin Boundary Contrast for the Non-Column and Non-Symmetrical Laue Case
(61.14; 61.70)
(a) Vol. **104**, No. 2, p. 519—527 (1987).

I. G. MARGVELASHVILI and Z. K. SARALIDZE
*Specific Features of the Point Defect Equilibrium Distribution and Kinetics in the Ordered Two-Component (Non-Ionic) Crystals
(61.50 and 61.70)
(a) Vol. **104**, No. 2, p. 587—596 (1987).

T. TSUCHIDA and R. KANEKO
*Temperature Dependence of Activation Energy in Stage II Recovery in Deformed Aluminium
(61.70; 72.15; S4)
(a) Vol. **104**, No. 2, p. 627—634 (1987).

A. JAKUBOWICZ, H.-U. HABERMEIER, A. EISENHEISS, and D. KÄSS
*Electron Beam Induced Current versus Temperature Investigations of Localized Dislocations in Heat-Treated Czochralski Silicon
(61.70; S5.11)
(a) Vol. **104**, No. 2, p. 635—641 (1987).

A. N. ISOTOV and E. A. STEINMAN
*Reconstruction of Optical Dislocation Centres under the Action of Shear Stresses
(61.70 and 71.55; 78.55; S5.11)
(a) Vol. **104**, No. 2, p. 777—784 (1987).

X. H. WANG, Y. B. XU, Z. G. WANG, and K. F. HA
*The Observation of the Dislocation-Free Zone ahead of the Crack-Tip in Bulk Aluminium Single Crystal
(61.70 and 62.20; S4)
(a) Vol. **104**, No. 2, p. K79—K82 (1987).

J. PAJAK and B. ROZENFELD
*Helium in Plastically Deformed Ni Samples Investigated by Means of the Positron Annihilation Method
(61.70 and 78.70; S1.2)
(a) Vol. **104**, No. 2, p. K85—K88 (1987).

61.80. Radiation damage and other irradiation effects

(for techniques of structure determination, see 61.10 to 61.16; for electron and ion impact phenomena, see 79.20)

B. RAUSCHENBACH and V. HEERA
*Ion Irradiation Induced Reamorphization of Boron Preimplanted Iron
(61.40 and 61.80; S1.1)
(a) Vol. **100**, No. 2, p. 423—429 (1987).

A. L. ASEEV, V. M. IVAKHNISHIN, D. HOEHL, and H. BARTSCH
Studies of the Interaction of Dislocations with Point Defects in Germanium by Means of Irradiation in a High-Voltage Electron Microscope
(61.70; 61.80; S5.12)
(a) Vol. **100**, No. 2, p. 431—440 (1987).

A. P. KIRILYUK
*Quantum Theory of Channeling in Deformed Crystals
(61.80)
(b) Vol. **141**, No. 1, p. 47—60 (1987).

P. K. J. BALAGARI and A. P. PATHAK
*A Model Potential Calculation for Positron Axial Channeling from Dislocation Affected Crystals
(61.80)
(b) Vol. **141**, No. 1, p. K15—K20 (1987).

P. H. DIONISIO and I. J. R. BAUMVOL
The Influence of Tin Ion Implantation and Radiation-Enhanced Tin Diffusion on the Thermal Oxidation of a High-Carbon Steel
(66.30; 61.80; S1.1; S10)
(a) Vol. **101**, No. 1, p. 63—76 (1987).

S. KOPTA, R. HAJDUK, A. Z. HRYNKIEWICZ, S. ŁAZARSKI, E. MAYDELL-ONDRUSZ, and B. RAJCHEL
*Rutherford Backscattering, Coulomb Excitation, and X-Ray Production Studied in Channeling Conditions
(61.80; 78.70; S1)
(a) Vol. **101**, No. 1, p. 77—96 (1987).

K. T. ANTONOVA, G. SCHIRMER, V. A. YAKOVLEV, and G. N. ZHIZHIN
Influence of Thermal Annealing on the Optical Properties of Crystalline Quartz after Oxygen and Nitrogen Ion Irradiation
(78.20; 61.80; S10.1)
(a) Vol. **101**, No. 1, p. 247—252 (1987).

I. A. ABROYAN, M. V. KRUGLOV, and I. K. SOLOMIN
*Surface-Layer Swelling on Implantation of Hydrogen Ions into Silicon at Different Doping Concentrations
(61.80; 68.20; S5.11)
(a) Vol. **101**, No. 1, p. K5—K8 (1987).

T. M. ZHANTIKIN, A. K. ZHETBAEV, M. F. VERESHCHAK, and Z. T. SERIKBAEVA
*Mössbauer Investigations of Radiation Damage in Ni_3Fe
(61.80; 76.80; S1.1; S1.2)
(a) Vol. **101**, No. 2, p. 361—370 (1987).

MENG XIANG-TI, DU YONG-CHANG, and FAN YU-DIAN
New Si–H IR Bands in Neutron-Irradiated Silicon Grown in Hydrogen Atmosphere
(78.30; 61.80; S5.11)
(a) Vol. **101**, No. 2, p. 619—628 (1987).

SP. DEDOUSSIS, G. D. VASILLIADES, A. LIOLIOS, C. TOURAMANIS, M. CHARDALAS, C. A. ELEFTERIADES, and ST. CHARALAMBOUS
Positron Annihilation Studies in Ionic Crystals
(78.70; 61.80; S9.11; S9.12)
(a) Vol. **102**, No. 2, p. 577—581 (1987).

N. S. SALEH and K. A. AL-SALEH
*Measurement of K-Shell X-Ray Cross Sections of Selected Elements from Ti to Zn for Incident Protons
(61.80; 78.70; S1.1; S1.2)
(a) Vol. **102**, No. 2, p. 619—623 (1987).

K. HOLLDACK, H. KERKOW, and M. GERICKE
*Rate Effects during Radiation-Enhanced Diffusion of Boron in Silicon
(61.80 and 66.30; S5.11)
(a) Vol. **102**, No. 2, p. 633—638 (1987).

C. A. LONDOS
*Annealing Studies of Defects Pertinet to Radiation Damage in Si:B
(61.80; 71.55; S5.11)
(a) Vol. **102**, No. 2, p. 639—644 (1987).

B. RAUSCHENBACH
*The Characteristic Temperature of Ion Beam Mixing
(61.80; 79.20; S1; S4)
(a) Vol. **102**, No. 2, p. 645—651 (1987).

P. V. KUCHINSKII and V. M. LOMAKO
*On the Mechanism of Primary Radiation Defect Annihilation in Si
(61.80; S5.11)
(a) Vol. **102**, No. 2, p. 653—658 (1987).

G. J. PAPAIOANNOU
*Electron Radiation Effects on GaAs MESFETs
(61.80 and 85; 73.40; S7.12)
(a) Vol. **102**, No. 2, p. 843—848 (1987).

R. WÜRSCHUM and H.-E. SCHAEFER
Defects in Low-Temperature Electron-Irradiated GaAs Studied by Positrons
(61.70; 61.80; 78.70; S7.12)
(a) Vol. **103**, No. 1, p. 101—105 (1987).

D. N. SEIDMAN, R. S. AVERBACK, and R. BENEDEK
*Displacement Cascades. Dynamics and Atomic Structure
(61.80; 61.70; S1; S1.2)
(b) Vol. **144**, No. 1, p. 85—104 (1987).

R. URBAN, P. EHRHART, W. SCHILLING, H. R. SCHOBER, and H. LAUTER
*Investigation of the Resonant Vibration Modes of Radiation Induced Interstitial Atoms in Aluminum by Diffuse Inelastic Neutron Scattering
(61.80 and 63.20; S4)
(b) Vol. **144**, No. 1, p. 287—304 (1987).

R. C. MARTIN and N. M. GHONIEM
*Monte Carlo Simulation of Coupled Ion-Electron Transport in Semiconductors
(61.80 and 73.40; 79.20; S5.11)
(a) Vol. **104**, No. 2, p. 743—754 (1987).

H. JOHANSEN and B. LÄMMEL
*Optical Feedback during CW Laser annealing of Silicon Detected by SEM-EBIC
(61.80; 79.20; S5.11)
(a) Vol. **104**, No. 2, p. K83—K84 (1987).

62. Mechanical and acoustical properties of condensed matter
(see also 61.70. Defects in crystals, 68.30. Surfaces and interfaces)

62.20. Mechanical properties of solids
(related to microscopic structure)

P. Gondi, A. Sili, and E. Bonetti
Relaxation Phenomena of Dislocations Contributing to Deformation of Al at Intermediate Temperatures
(61.70; 62.20; 62.40; S4)
(a) Vol. **99**, No. 2, p. 375—384 (1987).

O. B. Arkan and H. Neuhäuser
Dislocation Velocities in Cu–Ni Alloys Determined by the Stress Pulse–Etch Pit Technique and by Slip Line Cinematography
(61.70; 62.20; S1.2)
(a) Vol. **99**, No. 2, p. 385—397 (1987).

F. Povolo and R. E. Bolmaro
*Average Elastic Constants and Tensor Invariants
(62.20; S1.2; S3; S11.1; S12)
(a) Vol. **99**, No. 2, p. 423—436 (1987).

A. G. Ivanchin
*An Application of Potential Theory for Solving the Plane Elastic Crack Problem
(62.20)
(a) Vol. **99**, No. 2, p. 437—447 (1987).

F. Wallow, G. Neite, W. Schröer, and E. Nembach
*Stiffness Constants, Dislocation Line Energies, and Tensions of Ni_3Al and of the γ'-Phases of NIMONIC 105 and of NIMONIC PE16
(62.20; 62.30; S1.2)
(a) Vol. **99**, No. 2, p. 483—490 (1987).

H.-J. Weiss
*On Deriving Vickers Hardness from Penetration Depth
(62.20)
(a) Vol. **99**, No. 2, p. 491—501 (1987).

M. Kumari and N. Dass
*Pressure Dependence of the Bulk Modulus in Solids
(62.20)
(a) Vol. **99**, No. 2, p. K75—K77 (1987).

P. Will and B. Michel
*Transformation Toughening Fatigue Crack Growth
(62.20)
(a) Vol. **99**, No. 2, p. K79—K82 (1987).

G. M. S. Srivastava, R. K. Sinha, and J. Shanker
*Interrelationship between Higher-Order Elastic Constants and Pressure Derivatives of Dielectric Constants for Ionic Crystals
(62.20 and 77.20; S9.11)
(b) Vol. **140**, No. 1, p. 89—95 (1987).

M. T. Sprackling and M. E. Bahrololoom
The Role of Iodide in the Strain-Ageing of Silver Iodobromide
(61.70; 62.20; S9.15)
(a) Vol. **100**, No. 1, p. 105—108 (1987).

R. Lacroix
Spin Hamiltonian of Cr^{3+} in a Strained Ruby Crystal
(71.70; 62.20; S10.1)
(b) Vol. **141**,-No. 2, p. 567—573 (1987).

I. V. Bodnar, B. V. Korzun, and A. P. Chernyakova
*Microhardness of the $A^{I}B^{III}C_2^{VI}$ Ternary Semiconductors and Their Solid Solutions
(62.20; S8.16)
(a) Vol. **101**, No. 2, p. 409—419 (1987).

I. M. Silvestrova, Y. V. Pisarevskii, P. A. Senyushenkov, A. I. Krupny, R. Voszka, I. Földvári, and J. Janszky
*Temperature Dependence of Elastic Properties of Paratellurite
(62.20; 62.65; 65.70; S10.1)
(a) Vol. **101**, No. 2, p. 437—444 (1987).

A. H. Phillips and N. K. Gobran
*Computation of the Stress Exponent of Cobalt in Its H.C.P. and F.C.C. Phases
(62.20; S1.2)
(a) Vol. **101**, No. 2, p. K115—K119 (1987).

V. Ramamurthy
On the Inconsistencies of Axially Symmetric Force Models for Noncubic Solids
(63.20; 62.20; S4)
(b) Vol. **142**, No. 1, p. 67—76 (1987).

H.-Matsuo Kagaya and T. Soma
Mode Grüneisen Parameters and Pressure-Induced Phase Transition of Tetrahedral Compounds
(63.20; 62.20; 64.70; S7.1)
(b) Vol. **142**, No. 1, p. 97—103 (1987).

G. Brauer and K. Popp
Neutron Embrittlement of Reactor Pressure Vessel Steels: A Challenge to Positron Annihilation and Other Methods
(61.80 and 78.70; 62.20; S1.1)
(a) Vol. **102**, No. 1, p. 79—90 (1987).

K. Ghazi-Wakili, U. Zimmermann, J. Brunner, P. Tipping, W. G. Waeber, and F. Heinrich
Positron Annihilation Studies on Neutron Irradiated Pressure Vessel Steels
(61.80 and 78.70; 62.20; 64.75; S1.1)
(a) Vol. **102**, No. 1, p. 153—163 (1987).

D. O. Northwood and I. O. Smith
*The Creep Behaviour of a Neutron Irradiated Al–0.3 wt% Li Alloy at 300 °C
(61.70 and 62.20; S4)
(a) Vol. **102**, No. 1, p. 251—263 (1987).

T. H. Youssef, F. A. Saadalah, and I. K. Bishay
*Plastic Behaviour of Polycrystalline Al–0.72 wt% Si
(62.20; S4)
(a) Vol. **102**, No. 1, p. 273—278 (1987).

J. Kumar, M. Thirumavalavan, F. D. Gnanam, and P. Ramasamy
*Microindentation Studies on Single Crystals of Sodium and Ammonium Fluoroborates
(62.20; S9.16)
(a) Vol. **103**, No. 2, p. 431—434 (1987).

Y. Nakada and T. Imura
*Influence of Dopants and Deformation Temperature on the Strength of GaAs
(62.20; 61.70; S7.12)
(a) Vol. **103**, No. 2, p. 435—442 (1987).

S. Ohtaki
*The Nature of Stage II_A Recovery in Cold-Worked Aluminium
(62.20; 61.70; 72.15; S4)
(a) Vol. **103**, No. 2, p. 449—457 (1987).

C. Ascheron, H. Neumann, V. N. Bugrov, and S. A. Karamyan
*Microhardness of Ion Implanted GaP
(61.70 and 62.20; S7.11)
(a) Vol. **103**, No. 2, p. K81—K84 (1987).

M. Ohta, T. Kanadani, A. Sakakibara, and M. Yamada
*On the Two Populations of GP Zones in the Al–Zn Alloy Aged after Two-Step Quenching
(62.20; S4)
(a) Vol. **103**, No. 2, p. K85—K86 (1987).

E. Kröner
*Stress Space and Strain Space in Continuum Mechanics
(62.20)
(b) Vol. **144**, No. 1, p. 39—44 (1987).

H. Günther
*On Continuum Mechanics and Special Relativity
(62.20)
(b) Vol. **144**, No. 1, p. 45—48 (1987).

M. Condat and H. O. K. Kirchner
*Computational Anisotropic Elasticity
(62.20)
(b) Vol. **144**, No. 1, p. 137—143 (1987).

F. R. N. Nabarro
*Plastic Deformation in Tension and in Torsion
(62.20; 61.70)
(a) Vol. **104**, No. 1, p. 47—50 (1987).

M. Werner
*Temperature and Strain-Rate Dependence of the Flow Stress of Ultrapure Tantalum Single Crystals
(62.20; 61.70; S1)
(a) Vol. **104**, No. 1, p. 63—78 (1987).

K. P. THAKUR and S. AHMAD
Phase Transformation in F.C.C. Metals and Their Stability on the Path of Transformation
(64.60; 62.20; S1; S1.2; S4)
(b) Vol. **144**, No. 2, p. 529—536 (1987).

P. KRATOCHVÍL, M. MANDULA, J. MENCL, J. PEŠIČKA, and B. SMOLA
The Constitution of the Cu–9 Ni–6 Sn Alloy Determined by Changes of the Strength and by Structure Determination during Annealing
(61.55; 62.20; S1.2)
(a) Vol. **104**, No. 2, p. 597—602 (1987).

E. BORCHI, S. DE GENNARO, L. LOMBARDINI, and M. ZOLI
*Elastic Constants and Young's Modulus for Texturized Materials with Cubic and Hexagonal Symmetries
(62.20; S1.2; S4)
(a) Vol. **104**, No. 2, p. 667—678 (1987).

X. H. WANG, Y. B. XU, Z. G. WANG, and K. F. HA
*The Observation of the Dislocation-Free Zone ahead of the Crack-Tip in Bulk Aluminium Single Crystal
(61.70 and 62.20; S4)
(a) Vol. **104**, No. 2, p. K79—K82 (1987).

B. VENGATESAN, N. KANNIAH, and P. RAMASAMY
*CVT Growth, Microhardness, and Electrical Characterization of $ZnGa_2S_4$ Single Crystals
(62.20 and 72.80; S8.16)
(a) Vol. **104**, No. 2, p. K93—K96 (1987).

O. KOHMOTO
*Microhardness of Amorphous Co-Rich FeCo–SiB Alloys
(62.20; S1.2)
(a) Vol. **104**, No. 2, p. K101—K103 (1987).

62.30. Mechanical and elastic waves

F. WALLOW, G. NEITE, W. SCHRÖER, and E. NEMBACH
Stiffness Constants, Dislocation Line Energies, and Tensions of Ni_3Al and of the γ'-Phases of NIMONIC 105 and of NIMONIC PE16
(62.20; 62.30; S1.2)
(a) Vol. **99**, No. 2, p. 483—490 (1987).

A. SZCZEPAŃSKI
*On the Radiation Process in Incompressible Elastic Media
(62.30)
(b) Vol. **142**, No. 2, p. 379—385 (1987).

S. A. GUNDERSEN and J. LOTHE
*A New Method for Numerical Calculations in Anisotropic Elasticity Problems
(62.30; 61.70)
(b) Vol. **143**, No. 1, p. 73—85 (1987).

62.40. Anelasticity, internal friction, and mechanical resonances

For thermomechanical effects, see 65.70
For magnetomechanical effects, see 75.80
For piezoelectric effects, see 77.60
For elasto-optical effects, see 78.20

HÈ YIZHEN and LI XIAO-GUANG
*A New Type of Internal Friction Peak of Metallic Glasses near T_g
(62.40; S1.2; S1.61)
(a) Vol. **99**, No. 1, p. 115—120 (1987).

E. K. NAIMI, N. A. TYAPUNINA, and G. Z. KURBANOV
Dislocation Internal Friction in CsI Single Crystals
(61.70; 62.40; S9.11)
(a) Vol. **99**, No. 2, p. 353—363 (1987).

P. GONDI, A. SILI, and E. BONETTI
Relaxation Phenomena of Dislocations Contributing to Deformation of Al at Intermediate Temperatures
(61.70; 62.20; 62.40; S4)
(a) Vol. **99**, No. 2, p. 375—384 (1987).

K. MIURA
*Nonlinear Resonances in Mn–7 at% Cu
(62.40; 64.70; S1.2)
(a) Vol. **99**, No. 2, p. 475—482 (1987).

M. GABBAY, A. VINCENT, and G. FANTOZZI
*Amplitude-Dependent Internal Friction Due to Thermomechanical Breakaway of Dislocations from Pinning Point Defects
(62.40; 61.70)
(a) Vol. **100**, No. 1, p. 121—137 (1987).

K. IWASAKI
Strain Amplitude Dependence of the Grain Boundary Internal Friction Peak of Dilute Aluminium Alloys
(61.70; 62.40; S4)
(a) Vol. **100**, No. 2, p. 453—459 (1987).

K. IWASAKI
Effects of Annealing and Tensile Stress on the Grain Boundary Internal Friction Peak of Pure Aluminum
(61.70; 62.40; S4)
(a) Vol. **101**, No. 1, p. 97—104 (1987).

WANG YENING, SUN WENYUAN, CHEN XIAOHUA, SHEN HUIMIN, and LU BAOSHENG
*Internal Friction Associated with the Domain Walls and the Second-Order Ferroelastic Transition in LNPP
(62.40; 64.70; S11)
(a) Vol. **102**, No. 1, p. 279—285 (1987).

T. S. KÊ, Q. TAN, and Q. F. FANG
*Further Experiments on the Anomalously Amplitude-Dependent Internal Friction Peaks in Polycrystalline and Single-Crystal Al–Mg
(62.40; S4)
(a) Vol. **103**, No. 2, p. 421—429 (1987).

E. MANN
*Eigenstates of Kinked Dislocations in Crystals
(61.70 and 62.40; 63.90)
(b) Vol. **144**, No. 1, p. 115—127 (1987).

B. S. BERRY and W. C. PRITCHET
*Computer Modeling of Relaxation Peaks Involving a Gaussian Distribution of Activation Energies
(62.40 and 77.40)
(b) Vol. **144**, No. 1, p. 375—384 (1987).

W. BENOIT, M. BUJARD, and G. GREMAUD
Kink Dynamics in F.C.C. Metals
(61.70; 62.20; 62.40; 62.80; S4)
(a) Vol. **104**, No. 1, p. 427—441 (1987).

H. INDRAWIRAWAN, O. BUCK, and O. N. CARLSON
*Substitutional–Interstitial Solute Interactions in Niobium–Vanadium–Oxygen Alloys
(62.40; 61.70; 66.30; S1; S1.2)
(a) Vol. **104**, No. 1, p. 443—451 (1987).

N. KURAMOCHI, H. MIZUBAYASHI, and S. OKUDA
Low-Temperature Dislocation Internal Friction in Hydrogen-Free Niobium
(61.70; 62.40; S1)
(a) Vol. **104**, No. 1, p. 453—467 (1987).

H. MIZUBAYASHI, Y. KATOH, and S. OKUDA
*Hydrogen Internal Friction Peak and the Effect of Quenching in Amorphous CuTi
(61.40 and 62.40; S1.2)
(a) Vol. **104**, No. 1, p. 469—485 (1987).

B. K. KARDASHEV, H.-J. KAUFMANN, A. B. LEBEDEV, and P. P. PAL-VAL
*On the Dislocation Internal Friction in High-Purity Molybdenum Single Crystals
(61.70 and 62.40; S1)
(a) Vol. **104**, No. 1, p. K17—K20 (1987).

F. POVOLO
*Some Considerations about Internal Friction Peaks Described by a Single Relaxation Time
(62.40)
(a) Vol. **104**, No. 2, p. 693—705 (1987).

S. C. YAN and T. S. KÊ
*Internal Friction Peaks Associated with the Polygonization Boundaries in Aluminium and Dilute Aluminium–Copper Alloys
(62.40; S4)
(a) Vol. **104**, No. 2, p. 715—721 (1987).

Q. TAN and T. S. KÊ
*Double Amplitude Internal Friction Peaks in Al–0.02 wt% Mg Single Crystals
(62.40; S4)
(a) Vol. **104**, No. 2, p. 723—729 (1987).

D. LASZIG and P. HAASEN
*Internal Friction of GaAs with EL2 Defects
(62.40; 71.55; S7.12)
(a) Vol. **104**, No. 2, p. K105—K109 (1987).

62.50. High-pressure and shock-wave effects in solids

62.60. Acoustical properties of liquids

For lattice dynamics, phonons, see 63
For second sound in quantum fluids, see 67.40

62.65. Acoustical properties of solids

For magnetoacoustic effects in metals, see 75.80
For acoustoelectric effects, see 72.50
For acousto-optical effects, see 78.20

F. FEDOROV
*On One Incorrect Approach in the Elastic Waves Theory
(62.65)
(b) Vol. **141**, No. 1, p. K25—K28 (1987).

I. M. SILVESTROVA, Y. V. PISAREVSKII, P. A. SENYUSHENKOV, A. I. KRUPNY, R. VOSZKA, I. FÖLDVÁRI, and J. JANSZKY
Temperature Dependence of Elastic Properties of Paratellurite
(62.20; 62.65; 65.70; S10.1)
(a) Vol. **101**, No. 2, p. 437—444 (1987).

62.80. Ultrasonic relaxation

(see also 74.30. Ultrasonic attenuation in superconductors)

G. G. BOKUCHAVA and A. I. TUGUSHI
*Ultrasonic Absorption in Hydrogen Containing Metals
(62.80; 61.70)
(b) Vol. **142**, No. 1, p. K1—K6 (1987).

MOHEB H. YOUSSEF and M. ROUSHDY
*A Modified Method for Measuring Ultrasonic Attenuation in Highly Attenuating Materials
(62.80; S1.2)
(a) Vol. **102**, No. 1, p. 287—292 (1987).

W. BENOIT, M. BUJARD, and G. GREMAUD
Kink Dynamics in F.C.C. Metals
(61.70; 62.20; 62.40; 62.80; S4)
(a) Vol. **104**, No. 1, p. 427—441 (1987).

63. Lattice dynamics and crystal statistics

(see also 65. Thermal properties, 66.70. Thermal conduction, 68.30. Dynamics of surface and interface vibrations, 78.30. Infrared and Raman spectra)

63.10. General theory

S. K. Bose and U. B. Dubey
*Frequencies of a General Damped Coupled Nonlinear Oscillator System
(63.10)
(b) Vol. **140**, No. 2, p. 401—406 (1987).

U. Lindner
To Applicability of Hamilton's Equations in Quantum Many-Body Problems
(75.10; 63.10; 71.10)
(b) Vol. **141**, No. 1, p. 263—268 (1987).

J. M. Wesselinowa
*Temperature and Anharmonicity Dependence of the Biphonon Linewidth
(63.10 and 63.20)
(b) Vol. **142**, No. 2, p. K113—K117 (1987).

H. Böttger, A. Freyberg, and D. Wegener
*Recursion Method for the Density of States of Bond-Diluted Central-Force Elastic Triangular Lattices
(63.10)
(b) Vol. **143**, No. 1, p. 105—111 (1987).

63.20. Phonons and vibrations in crystal lattices

A. Ghorai and S. K. Sen
Calculations on Relaxations Due to a Vacancy, Binding Energy of a Vacancy–Vacancy Pair, and Formation Energy of a Vacancy in Thorium
(61.70; 63.20; S1.5)
(b) Vol. **139**, No. 1, p. 77—83 (1987).

W. Y. Zeng, J. F. Higgs, and A. Anderson
*Lattice Dynamics of Tin Tetrabromide
(63.20; S9)
(b) Vol. **139**, No. 1, p. 85—96 (1987).

M. Klanjšek Gunde, B. Orel, and V. Moiseenko
*Vibrational Studies of and Impurity Mode Behaviour in Mixed $Pb_5(GeO_4)_{1-x}(SiO_4)_x(VO_4)_2$ Monocrystals
(63.20; 78.30; S10.15)
(b) Vol. **139**, No. 1, p. K75—K80 (1987).

A. MEMON and D. B. TANNER
*Phonon Combination Bands in the Far-Infrared Spectrum of $K_{0.5}Rb_{0.5}I$ Mixed Crystals
(63.20 and 78.30; S9.15)
(b) Vol. **139**, No. 1, p. K81—K85 (1987).

J. FONTCUBERTA
*Lattice Dynamics Study of Polycrystalline ^{57}CoO by Mössbauer Spectroscopy
(63.20; 76.80; S10.1)
(b) Vol. **139**, No. 2, p. 379—386 (1987).

J. TATARKIEWICZ and K. WINER
*Vibrational Properties of Hydrogen in Proton-Implanted Silicon
(61.70 and 63.20; 78.30; S5.11)
(b) Vol. **139**, No. 2, p. 409—415 (1987).

H.-MATSUO KAGAYA, N. SHOJI, and T. SOMA
Lattice Vibrational Contributions to the Bulk Properties of Si and Ge
(62.20; 63.20; S5.11; S5.12)
(b) Vol. **139**, No. 2, p. 417—422 (1987).

M. KLANIŠEK GUNDE, B. OREL, and V. MOISEENKO
*Infrared Reflection Spectra of and Polariton Dispersion in Ferroelectric $Pb_5Ge_3O_{11}$
(63.20 and 78.30; S10)
(b) Vol. **139**, No. 2, p. K151—K155 (1987).

E. SCHULTHEISS, A. SCHARMANN, and D. SCHWABE
Energy Levels of Triply Ionized Praseodymium in $BiLiF_4$ Crystals
(71.70 and 78.40; 63.20; S9.16)
(b) Vol. **140**, No. 1, p. 173—189 (1987).

J. TATARKIEWICZ, A. KRÓL, A. BREITSCHWERDT, and M. CARDONA
*Localized Vibrations of Hydrogen and Deuterium in GaAs. A Comparison with Ge and Si
(63.20; 61.70; 78.50; S7.12)
(b) Vol. **140**, No. 2, p. 369—375 (1987).

B. I. CRAIG and P. V. SMITH
A Semi-Empirical Approach to the Bulk Properties of Copper
(71.10 and 71.25; 62.20; 63.20; S1.2)
(b) Vol. **140**, No. 2, p. 501—508 (1987).

V. ŽELEZNÝ, J. PETZELT, and R. SWIETLIK
Infrared Properties of $TeEA(TCNQ)_2$
(78.30; 63.20; 71.38; S12.1)
(b) Vol. **140**, No. 2, p. 595—604 (1987).

E. CAZZANELLI, A. FONTANA, F. RÓCCA, and R. FRECH
Low-Temperature Behaviour of the Lithium Modes in Monoclinic Li_2SO_4
(78.30; 63.20; S11.1)
(b) Vol. **140**, No. 2, p. 621—630 (1987).

H.-Matsuo Kagaya and T. Soma
*Mode Grüneisen Parameters and Pressure-Induced Phase Transition of Tetrahedral Compounds
(63.20; 62.20; 64.70; S7.1)
(b) Vol. **142**, No. 1, p. 97—103 (1987).

J. Otto
*Investigation of the Phonon Sidebands of SH^- and SeH^- Centres in Cesium Halides and Potassium Bromide
(63.20 and 78.30; S9.11)
(b) Vol. **142**, No. 1, p. 105—115 (1987).

J. Monecke, W. Cordts, G. Irmer, B. H. Bairamov, and V. V. Toporov
*LO-Phonon Modes Bound to Neutral Impurities in Planar Semiconductors. II. Special Cases and Experimental Results
(63.20 and 71.55; 71.38; S7.11)
(b) Vol. **142**, No. 1, p. 237—246 (1987).

Xichen Yang, Guoxiang Lan, Bing Li, and Huafu Wang
*Raman Spectra and Directional Dispersion in $LiNbO_3$ and $LiTaO_3$
(78.30; 63.20; S11.1)
(b) Vol. **142**, No. 1, p. 287—300 (1987).

V. K. Thakur and T. N. Singh
*Lattice Dynamics of Al and Pb
(63.20; 65.40; S4)
(b) Vol. **142**, No. 2, p. 401—406 (1987).

J. Tatarkiewicz, A. M. Witowski, and E. Jahne
*Hydrogen Vibrations in CdTe
(61.70 and 63.20; 78.30; S8.13)
(b) Vol. **142**, No. 2, p. K101—K104 (1987).

J. C. Upadhyaya and O. P. Kulshrestha
*Lattice Dynamics of Ca in the Resonant Model Potential
(63.20; S3)
(b) Vol. **142**, No. 2, p. K109—K112 (1987).

J. M. Wesselinowa
*Temperature and Anharmonicity Dependence of the Biphonon Linewidth
(63.10 and 63.20)
(b) Vol. **142**, No. 2, p. K113—K117 (1987).

C. Malinowska-Adamska
*Self-Consistent Calculations of the Melting Criterion Based on Lattice Instability
(63.20 and 64.70; S1.2; S1.3; S2; S4)
(b) Vol. **143**, No. 1, p. 63—71 (1987).

D. W. Field
*Electronic Transfer and Thermal Vibration Parameters in Titanium Carbide. An X-Ray Diffraction Study
(63.20; 61.50; S1.61)
(b) Vol. **143**, No. 1, p. 87—90 (1987).

V. P. Gribkowski, A. L. Gurskii, G. A. Pashkevich, and G. P. Yablonskii
Crystallographic Orientation of Electric Discharges in TeO_2 and $LiNbO_3$ Monocrystals
(77.50; 63.20; S10.1; S11.1)
(a) Vol. **103**, No. 2, p. K153—K156 (1987).

R. Urban, P. Ehrhart, W. Schilling, H. R. Schober, and H. Lauter
*Investigation of the Resonant Vibration Modes of Radiation Induced Interstitial Atoms in Aluminium by Diffuse Inelastic Neutron Scattering
(61.80 and 63.20; S4)
(b) Vol. **144**, No. 1, p. 287—304 (1987).

E. A. Vinogradov, N. M. Gasanly, L. G. Gasanova, A. Z. Magomedov, and V. I. Tagirov
Infrared Reflectivity Spectra of $Cu_3In_5S_9$ Layer Single Crystals
(78.30; 63.20; S8.16)
(b) Vol. **144**, No. 1, p. K73—K76 (1987).

H. He
*Electronic and Vibrational Properties of One-Dimensional Quasi-Crystals
(63.20 and 71.20; S12.1)
(b) Vol. **144**, No. 2, p. 631—638 (1987).

U. Döbler and H. Happ
Infrared Investigation of Phase Transitions in $CsHSO_4$
(78.30; 63.20; 77.80; S11.1)
(b) Vol. **144**, No. 2, p. 853—862 (1987).

P. Gard, C. Sourisseau, and O. Gorochov
Effect of Metal Substitution in ZrS_3 and TiS_3 Compounds. Electronic, Raman, and Resonance Raman Study of $Zr_{1-x}Ti_xS_3$ Ternary Phass, $0 < x < 1$
(78.30; 63.20; S8.16)
(b) Vol. **144**, No. 2, p. 885—901 (1987).

T. I. Galkina, A. Yu. Blinow, M. M. Bonch-Osmolovskii, P. N. Lebedev, O. Koblinger, K. Lassmann, and W. Eisenmenger
*Down-Conversion of High-Frequency Acoustic Phonons in a-Si:H
(63.20; S5.11)
(b) Vol. **144**, No. 2, p. K87—K91 (1987).

G. J. Vázquez and L. F. Magaña
Ab-initio Calculation of the Volume Dependence of the Resistivity of Aluminium
(72.15; 63.20; S4)
(b) Vol. **144**, No. 2, p. K119—K123 (1987).

M. Jouanne, C. Julien, and M. Balkanski
*Polarization Studies of Polar Phonons of InSe
(63.20 and 78.20; S8)
(b) Vol. **144**, No. 2, p. K147—K150 (1987).

63.50. Vibrational states in disordered systems

K. K. Mamedov, A. B. Abdullaev, B. Z. Shalumov, M. I. Mekhtiev, M. A. Alyanov, and D. O. Gumbatov
Low-Temperature Heat Capacity and Thermodynamic Properties of Silicon-Dioxide-Based Binary Vitreous Systems
(65.40 and 65.50; 63.50; S10.15)
(a) Vol. **99**, No. 2, p. 413—421 (1987).

I. Bratu, P. Oană, and M. Culea
An Infrared Study of $x\%$ BaO · 6 Fe_2O_3 · $(100 - x)\%$ 3 B_2O_3 · 2 BaO Glasses
(78.30; 63.50; S10.15)
(a) Vol. **100**, No. 2, p. K195—K198 (1987).

J. Monecke
Material Constants of Composites. II. An Improvement over the Coherent Potential Approximation for Crystals with Site-Diagonal Disorder
(71.10; 63.50)
(b) Vol. **141**, No. 1, p. 173—176 (1987).

A. Yu. Rumyantsev, N. L. Mitrofanov, A. S. Ivanov, N. A. Chernoplekov, L. Weiss, H. Eschrig, and R. Richter
*Anomalies in the Phonon Spectra and the Fermi Surface of the Random Substitutional Alloy $Cu_{0.9}Ni_{0.1}$
(63.50 and 71.25; S1.2)
(b) Vol. **143**, No. 2, p. 625—636 (1987).

N. M. Plakida, S. Stamenković, and G. M. Vujičić
*The Model Description of Neutron Scattering by Quasilocal Vibrations and Phonons in Glasses
(63.50)
(b) Vol. **144**, No. 2, p. 537—542 (1987).

63.70. Statistical mechanics of lattice vibrations

(see also 65. Thermal properties of condensed matter, and 66.70. Thermal conduction)

A. Holz
*Gauge Theory of Defect Systems and Equivalence Principle
(61.70 and 63.70)
(b) Vol. **144**, No. 1, p. 49—71 (1987).

63.90. Other topics in lattice dynamics and crystal statistics

K. VENU, V. S. S. SASTRY, and J. RAMAKRISHNA
Molecular Dynamics in $[(CH_3)_4N]_2CdI_4$. A Proton Magnetic Relaxation Study
(76.60; 63.90; S9.16)
(b) Vol. **140**, No. 1, p. 251—259 (1987).

F. A. KHWAJA, S. H. NAQVI, and M. S. K. RAZMI
*A Comparative Study of the Phenomenological Potentials for Ionic Crystals of Alkali Halides
(61.50 and 63.90; S9.11)
(b) Vol. **143**, No. 2, p. 453—462 (1987).

E. MANN
Eigenstates of Kinked Dislocations in Crystals
(61.70 and 62.40; 63.90)
(b) Vol. **144**, No. 1, p. 115—127 (1987).

64. Equations of state, phase equilibria, and phase transitions

64.10. General theory of equations of state and phase equilibria

64.30. Equations of state of specific substances

(see also 65.70. Thermal expansion)

64.60. General studies of phase transitions

(for magnetic, superconducting, and quantum fluid critical phenomena, see 75, 74, and 67, respectively)

I. A. Ovidko and A. E. Romanov
Topological Excitations (Defects, Solitons, Textures, Frustrations) in Condensed Media **(Review Article)**
(61.70; 64.60; 71.45; 75.30)
(a) Vol. **104**, No. 1, p. 13—45 (1987).

F. El-Kabbany, G. Said, S. Mahrous, and N. H. Taher
*Thermal Hysteresis in Isomorphic Divalent Ionic Nitrates $Ba(NO_3)_2$, $Pb(NO_3)_2$, and $Sr(NO_3)_2$
(64.60)
(a) Vol. **99**, No. 1, p. 105—113 (1987).

S. Waplak, J. Minge, and T. Krajewski
*EPR Studies of the Low-Temperature Phase Transition in $LiKSO_4$ by SO_4^- and CrO_4^{3-} Centres
(64.60 and 76.30; S11.1)
(a) Vol. **99**, No. 1, p. K27—K30 (1987).

J. Li and B. X. Liu
*CEMS to Characterize the Ion Induced Fe–Mo Metastable Phase
(64.60 and 76.80; S1; S1.1)
(a) Vol. **99**, No. 1, p. K31—K34 (1987).

G. Busiello and D. I. Uzunov
*On the Enhancement of Critical Fluctuations by Quenched Random Fields
(64.60)
(b) Vol. **139**, No. 2, p. K107—K112 (1987).

H. M. Tawancy and M. O. Aboelfotoh
*Experimental Observations of Transient Phases during Long-Range Ordering to Ni_4Mo in a Ni–Mo–Fe–Cr Alloys
(64.60; 61.14; S1.2)
(a) Vol. **99**, No. 2, p. 461—466 (1987).

A. I. Gusev and A. A. Rempel
*Calculating the Energy Parameters for the CV and OPF Methods
(61.50; 64.60; S1.61)
(b) Vol. **140**, No. 2, p. 335—346 (1987).

P. Négrier, N. B. Chanh, C. Courseille, C. Hauw, A. Méresse, and M. Couzi
*The Imma ↔ Pnma Phase Transition of $NH_3(CH_2)_5NH_3CdCl_4$ Studied by X-Ray Diffraction
(64.60; 61.10; S9)
(a) Vol. **100**, No. 2, p. 473—483 (1987).

J. Shanker, S. C. Sharma, and M. Kumar
*Analysis of the Structural Phase Transition in Mixed Alkali Halide Crystals
(64.60; S9.15)
(b) Vol. **141**, No. 2, p. 409—414 (1987).

M. Onoda and H. Nagasawa
*Phase Transition and Bipolaron Formation in Quasi-One-Dimensional Conductor β-$Ag_xV_2O_5$
(64.60 and 71.38; 76.30; 76.70; S11)
(b) Vol **141**, No. 2, p. 507—516 (1987).

64.70. Phase equilibria, phase transitions, and critical points of specific substances

H. Löffler, G. Wendrock, and K. Pawelzyk
The Influence of Small Mg Additions on the Stability of Guinier-Preston Zones in Al–4.5 at% Zn–x Mg Alloys
(61.55; 61.10; 64.70; S4)
(a) Vol. **99**, No. 1, p. 73—80 (1987).

S. Ahmed, A. M. Shamah, R. Kamel, and Y. Badr
*Structural Changes of $(NH_4)_2SO_4$ Crystals
(64.70; 77.80; S11.1)
(a) Vol. **99**, No. 1, p. 131—140 (1987).

R. A. Petrova
*Kinetics of Formation of the σ-Phase in Austenite-Ferrite Stainless Steel Type X18AГ12
(61.55 and 64.70; S1.1)
(a) Vol. **99**, No. 1, p. K1—K5 (1987).

M. Kumari, K. Kumari, and N. Dass
*On the Melting Law at High Pressures
(64.70; S2; S9)
(a) Vol. **99**, No. 1, p. K23—K26 (1987).

H.-Matsuo Kagaya and T. Soma
*The Pressure Effect on the Phase Diagrams at Low Temperatures of the Si–Ge System
(62.50 and 64.70; S5.11; S5.12)
(b) Vol. **139**, No. 2, p. K103—K106 (1987).

J. Mayer, K. Urban, and J. Fidler
*Phase Transitions between the Quasicrystalline, Crystalline, and Amorphous Phases in Al–16 at%V
(64.70; 61.40; S4)
(a) Vol. **99**, No. 2, p. 467—473 (1987).

K. Miura
Nonlinear Resonance in Mn–7 at%Cu
(62.40; 64.70; S1.2)
(a) Vol. **99**, No. 2, p. 475—482 (1987).

N. K. Sanyal, R. A. Yadav, and S. R. Shukla
Propagation of Ultrasonic Waves in a Smectic Liquid Crystal. Cholesteryl Erucate
(62.60; 64.70; S12)
(a) Vol. **100**, No. 1, p. 109—114 (1987).

A. M. Shamah, S. Ahmed, R. Kamel, and Y. Badr
*Structural Changes of $((NH_4)_{1-x}K_x)_2SO_4$ Crystals
(64.70; 61.60; S11.1)
(a) Vol. **100**, No. 1, p. 115—119 (1987).

L. Bosi
*A Simple Theory for Forecasting the Melting Temperature of Alkali Halides and of Alkaline-Earth Oxides
(64.70; 77.20; S9.11; S10.1)
(a) Vol. **101**, No. 2, p. K111—K114 (1987).

A. Wojakowska and E. Kundys
High Electrical Conductivity of a Solid Phase in the SnI_2–AgI System
(72.80; 64.70; S9.16)
(a) Vol. **101**, No. 2, p. K153—K155 (1987).

S. Abdi, Y. Mlik, and M. H. Ben Ghozlen
Group-Theoretical Analysis of the Symmetry Changes in R_2MX_4 Compounds
(61.50; 64.70; S9.16)
(b) Vol. **142**, No. 1, p. 27—34 (1987).

H.-Matsuo Kagaya and T. Soma
Mode Grüneisen Parameters and Pressure-Induced Phase Transition of Tetrahedral Compounds
(63.20; 62.20; 64.70; S7.1)
(b) Vol. **142**, No. 1, p. 97—103 (1987).

Xiao Keqin, Dong Yuanda, and He Yizhen
A Further Study on the Crystallization of a-$Pd_{80}Si_{20}$ and an Analysis of the Structure of the Amorphous Matrix
(61.40; 64.70; S1.3)
(a) Vol. **102**, No. 1, p. 221—229 (1987).

M. T. Sebastian, K. Narayanan, and P. Krishna
X-Ray Diffraction Effects from 2H Crystals Undergoing Transformation to the 3C Structure by the Layer Displacement Mechanism
(61.10; 61.70; 64.70; S6)
(a) Vol. **102**, No. 1, p. 241—249 (1987).

Wang Yening, Sun Wenyuan, Chen Xiaohua, Shen Huimin, and Lu Baosheng
Internal Friction Associated with the Domain Walls and the Second-Order Ferroelastic Transition in LNPP
(62.40; 64.70; S11)
(a) Vol. **102**, No. 1, p. 279—285 (1987).

B. T. Deshmukh, S. V. Bodade, and S. V. Moharil
Thermoluminescence and Reflectance Spectra of $K_2Ca_2(SO_4)_3$ and $K_2Mg_2(SO_4)_3$
(78.60; 64.70; S11.1)
(a) Vol. **102**, No. 1, p. 381—388 (1987).

Z. Jirák, E. Pollert, A. Tříska, and S. Vratislav
*Structure and the Orthorhombic–Tetragonal Transition in $YBa_2Cu_3O_7$
(61.60 and 64.70; S10)
(a) Vol. **102**, No. 2, p. K61—K66 (1987).

S. Ahmed, A. M. Shamah, R. Kamel, and Y. Badr
*Phase Transition in $[(NH_4)_{1-x}K_x]_2SO_4$ Crystals
(64.70; S11.1)
(a) Vol. **102**, No. 2, p. K81—K85 (1987).

64.75. Solubility, segregation, and mixing

V. I. DIMITROV and A. S. SHTEINBERG
*Short-Range Order in Multicomponent Solid Solutions
(64.75)
(b) Vol. **140**, No. 1, p. 83—88 (1987).

P. I. DIDENKO, A. A. EFREMOV, V. S. KHOMCHENKO, G. PH. ROMANOVA, and N. A. VLASENKO
Comparative Study of SiO_x and SiO_x:TbF_3 Films
(68.55; 64.75; 78.65; S10.1)
(a) Vol. **100**, No. 2, p. 501—511 (1987).

B. KLOTH, M. V. LUPAL, A. N. PIKHTIN, C. E. RICHTER, R. RIES, R. STEGMANN, and M. TRAPP
Determination of the Nitrogen Concentration in VPE-$GaAs_{1-x}P_x$
(68.55 and 78.65; 64.75; S7.15)
(a) Vol. **100**, No. 2, p. 545—552 (1987).

A. ZAIOUR, M. HAGE-ALI, J. M. KOEBEL, A. BENTZ, and P. SIFFERT
*Determination of Segregation Coefficients of a Series of Impurities in Tellurium
(61.70 and 64.75; S5)
(a) Vol. **100**, No. 2, p. K139—K143 (1987).

V. E. BORISENKO and S. G. YUDIN
*Steady-State Solubility of Substitutional Impurities in Silicon
(64.75; S5.11)
(a) Vol. **101**, No. 1, p. 123—127 (1987).

J. KIRCHHOF, P. KLEINERT, W. RADLOFF, and E. BELOW
Diffusion Processes in Lightguide Materials. The Diffusion of OH in Silica Glass at High Temperatures
(66.30; 64.75; S10.1)
(a) Vol. **101**, No. 2, p. 391—401 (1987).

H. PFEIFFER and R. HERGT
*Grain Boundary Segregation and Coercivity of Co/Cr Films for Perpendicular Magnetic Recording
(64.75 and 68.55; S1.2)
(a) Vol. **101**, No. 2, p. K121—K124 (1987).

A. BELAIDI, H. P. LEIGHLY, P. G. COLEMAN, and R. N. WEST
*A Positron Study of Defect Formation in Cd–Mg Alloys
(64.75 and 78.70; 61.70; S1)
(a) Vol. **102**, No. 1, p. 127—132 (1987).

W. LÜHR-TANCK, A. SAGER, M. EDERHOF, and H. BOSSE
*A Positron Annihilation Study of Precipitation Effects in <u>Cu</u>In and <u>Cu</u>Sb Alloys
(64.75 and 78.70; 61.70; S1.2)
(a) Vol. **102**, No. 1, p. 133—138 (1987).

64.80. Other phase properties of systems

M. Ben Salem

*Conditions générales d'obtention de parois cohérentes sans contraintes entre domaines élastiques
(General Conditions for Obtaining Coherent Walls under Strain between Elastic Domains)
(64.80; 68.48)
(a) Vol. **99**, No. 2, p. 449—459 (1987).

N. Willmann, W. Mader, E. Wachtel, and B. Predel

Decomposition and Crystallization Behaviour of the Amorphous $Ni_{39}Pd_{43}P_{18}$ and $Ni_{37}Pd_{45}P_{18}$ Alloys
(61.40; 64.80; 75.50; S1.2)
(a) Vol. **104**, No. 1, p. 369—379 (1987).

65. Thermal properties of condensed matter

(see also 63. Lattice dynamics; for thermodynamic properties of quantum fluids, see 67.40; for thermal properties of solid helium, see 67.80)

65.20. 65.20. Heat capacities of liquids

R. N. JOARDER and T. DAS
The Thermodynamic Properties of Liquid Transition and Rare-Earth Metals Based on CHS Model
(65.50; 65.20; S1.2; S1.4)
(b) Vol. **144**, No. 2, p. 557—564 (1987).

65.40. Heat capacities of solids

(for specific heat of superconductors, see 74.30; for specific heat of magnetic systems, see 75.40)

D. NEOGY and T. PUROHIT
The Behavior of Active Centers in a Laser Host. A Crystal Field Investigation of Sm^{3+} in LaF_3 Single Crystal
(71.70; 65.40; 75.30; S9)
(b) Vol. **139**, No. 2, p. 519—525 (1987).

K. K. MAMEDOV, A. B. ABDULLAEV, B. Z. SHALUMOV, M. I. MEKHTIEV, M. A. ALYANOV, and D. O. GUMBATOV
*Low-Temperature Heat Capacity and Thermodynamic Properties of Silicon-Dioxide-Based Binary Vitreous Systems
(65.40 and 65.50; 63.50; S10.15)
(a) Vol. **99**, No. 2, p. 413—421 (1987).

M. A. ALDZHANOV, N. G. GUSEINOV, and Z. N. MAMEDOV
*Heat Capacity of $TlGaSe_2$
(65.40; S8.16)
(a) Vol. **100**, No. 2, p. K145—K148 (1987).

G. F. CEROFOLINI and L. MEDA
*Low-Temperature Heat Capacity of Zero-Gap Semiconductors
(65.40)
(b) Vol. **141**, No. 2, p. K93—K95 (1987).

H.-MATSUO KAGAYA, N. SHOJI, and T. SOMA
*Specific Heat and Thermal Expansion at High Temperatures of Si and Ge
(65.40 and 65.70; S5.11; S5.12)
(b) Vol. **142**, No. 1, p. K13—K17 (1987).

V. K. THAKUR and T. N. SINGH
Lattice Dynamics of Al and Pb
(63.20; 65.40; S4)
(b) Vol. **142**, No. 2, p. 401—406 (1987).

H.-MATSUO KAGAYA and T. SOMA
*Thermal Properties of Tetrahedral Compounds
(65.40 and 65.70; S7.1; S8.1)
(b) Vol. **142**, No. 2, p. 411—416 (1987).

M. A. Gaffar, M. M. Mebed, and A. Abu El-Fadl
*The Specific Heat of Pure and Doped Triglycine Sulphate Single Crystals
(65.40 and 77.80; S12)
(a) Vol. **103**, No. 2, p. 459—466 (1987).

A. Kasten, H. G. Kahle, P. Klöfer, and D. Schäfer-Siebert
Specific Heat Experiments at the Cooperative Jahn-Teller Transition in the Mixed Crystal Systems $(Tb_x, Tm_{1-x})AsO_4$ and $(Tb_x, Tm_{1-x})VO_4$
(64.60; 65.40; S11.1)
(b) Vol. **144**, No. 1, p. 423—436 (1987).

H. Neumann, G. Kühn, and W. Möller
*Heat Capacity and Lattice Anharmonicity in Cu–III–VI_2 Chalcopyrite Compounds
(65.40; S8.16)
(b) Vol. **144**, No. 2, p. 565—573 (1987).

L. E. Shelimova and S. K. Plachkova
*Estimation of the Debye Temperature of IV–VI Semiconductor Compounds and Rhombohedral $(GeTe)_{1-x}((Ag_2Te)_{1-y}(Sb_2Te_3)_y)_x$ Solid Solutions ($y = 0.6$)
(65.40; 65.70; S8; S8.15)
(a) Vol. **104**, No. 2, p. 679—685 (1987).

65.50. Thermodynamic properties and entropy

S. V. Godoy, S. Fujita, and R. Rodriguez
Exact Solutions for Homogeneous Linear Biased Correlated Walks. A Case of the Boltzmann Entropy Decreasing in Time
(66.10 and 66.30; 65.50)
(b) Vol. **139**, No. 1, p. 67—76 (1987).

B. V. Egorov, I. B. Egorova, and M. A. Krivoglaz
The Electronic Heterogeneous States in Ferroelectric Semiconductors
(71.25 and 77.80; 65.50; 77.30)
(b) Vol. **139**, No. 1, p. 173—184 (1987).

K. K. Mamedov, A. B. Abdullaev, B. Z. Shalumov, M. I. Mekhtiev, M. A. Alyanov, and D. O. Gumbatov
*Low-Temperature Heat Capacity and Thermodynamic Properties of Silicon-Dioxide-Based Binary Vitreous Systems
(65.40 and 65.50; 63.50; S10.15)
(a) Vol. **99**, No. 2, p. 413—421 (1987).

L. M. Gratton, A. Miotello, and C. Tosello
Thermal Energy Balances in Ion Implanted Copper Bars
(61.80; 65.50; S1.2)
(a) Vol. **100**, No. 1, p. 53—57 (1987).

T. Bhattacharya and A. K. Ghosh
Pair-Correlation Functions of Liquid Noble Metals through Born-Green-Yvonne Theory
(61.25; 65.50; S1.2; S1.3)
(b) Vol. **140**, No. 2, p. 331—334 (1987).

Á. SZÁSZ
*The Exact Solution of the Real Square-Lattice-Gas System
(64.10 and 65.50)
(b) Vol. **140**, No. 2, p. 415—420 (1987).

U. F. EDGAL
*On the Free Energy of the Hard Sphere Fluid
(65.50)
(b) Vol. **141**, No. 2, p. 421—426 (1987).

K. N. KHANNA, C. PAL, and P. SINGH
*Volume of Mixing in Binary Liquid Alloys
(65.50; S2)
(b) Vol. **143**, No. 1, p. 55—61 (1987).

G. A. ABDURAGIMOV and R. P. MEILANOV
*Dynamics of the Temperature Field at Local Pulse Heat Release on a Solid Surface
(65.50)
(a) Vol. **104**, No. 1, p. K27—K31 (1987).

H.-MATSUO KAGAYA and T. SOMA
The Influence of the Compression Effect on the Phase Diagrams of Si–Ge System
(62.50; 64.75; 65.50; S6)
(b) Vol. **144**, No. 2, p. 549—556 (1987).

R. N. JOARDER and T. DAS
*The Thermodynamic Properties of Liquid Transition and Rare-Earth Metals Based on CHS Model
(65.50; 65.20; S1.2; S1.4)
(b) Vol. **144**, No. 2, p. 557—564 (1987).

65.70. Thermal expansion and thermomechanical effects

(see also 64.30. Equations of state)
For thermal conduction in nonmetallic liquids, see 66.60; for nonmetallic solids, see 66.70
For electronic thermal conduction, see 72.10 and 72.20
For thermal conductivity of superconductors, see 74.30
For pyroelectric and electrocalorix effects, see 77.70

N. S. ORLOVA and I. V. BONDAR
*The Anisotropy of the Thermal Expansion in the $AgInC_2^{VI}$ Chalcopyrite-Type Compounds from the X-Ray Diffraction Data
(65.70; S8.16)
(a) Vol. **101**, No. 2, p. 421—425 (1987).

I. M. SILVESTROVA, Y. V. PISAREVSKII, P. A. SENYUSHENKOV, A. I. KRUPNY, R. VOSZKA, I. FÖLDVÁRI, and J. JANSZKY
Temperature Dependence of Elastic Properties of Paratellurite
(62.20; 62.65; 65.70; S10.1)
(a) Vol. **101**, No. 2, p. 437—444 (1987).

H.-MATSUO KAGAYA, N. SHOJI, and T. SOMA
*Specific Heat and Thermal Expansion at High Temperatures of Si and Ge
(65.40 and 65.70; S5.11; S5.12)
(b) Vol. **142**, No. 1, p. K13—K17 (1987).

H.-MATSUO KAGAYA and T. SOMA
*Thermal Properties of Tetrahedral Compounds
(65.40 and 65.70; S7.1; S8.1)
(b) Vol. **142**, No. 2, p. 411—416 (1987).

T. SOMA, S. SUZUKI, and H.-MATSUO KAGAYA
Pressure Effect on the Melting Point of Noble Metals
(63.20; 65.70; S1.3)
(b) Vol. **143**, No. 1, p. 99—103 (1987).

P. ČAPKOVÁ, L. SMRČOK, V. ŠÍMA, V. VALVODA, B. CHALUPA, and P. ETTMAYER
Thermal Vibrations in Substoichiometric Vanadium Nitrides. Dynamical Deformation Effect
(63.20; 65.70; S1.62)
(b) Vol. **143**, No. 2, p. 471—476 (1987).

P. DEUS, H. A. SCHNEIDER, U. VOLAND, and K. STIEHLER
*Low Temperature Thermal Expansion of InP
(65.70; S7.11)
(a) Vol. **103**, No. 2, p. 443—447 (1987).

L. E. SHELIMOVA and S. K. PLACHKOVA
Estimation of the Debye Temperature of IV–VI Semiconductor-Compounds and Rhombohedral $(GeTe)_{1-x}((Ag_2Te)_{1-y}(Sb_2Te_3)_y)_x$ Solid Solutions ($y = 0.6$)
(65.40; 65.70; S8; S8.15)
(a) Vol. **104**, No. 2, p. 679—685 (1987).

65.90. Other topics in thermal properties of condensed matter

A. K. GEORGIEV, M. A. MISHEVA, and G. KH. TOUMBEV
*TDPAC Study of $HfO(NO_3)_2 \cdot 2\,H_2O$
(65.90 and 76.80; S11)
(a) Vol. **101**, No. 1, p. K19—K21 (1987).

P. WÜNSCHE, R. KUNZE, and M. WÄSCHE
*Activation Energy Distribution in Poly-(Methylmethacrylate–Methacrylic Acid) Estimated by EPR and Thermoluminescence
(65.90; 76.30; 78.60; S12)
(a) Vol. **103**, No. 1, p. 145—152 (1987).

66. Transport properties of condensed matter (nonelectronic)

66.10. Diffusion and ionic conduction in liquids

66.30. Diffusion in solids

V. V. Kolomytkin
*The Pipe Diffusion of the Self-Interstitial in a Loaded Crystal
(66.30; S1.1)
(a) Vol. **100**, No. 1, p. K17—K20 (1987).

D. Tsoukalas and P. Chenevier
*Boron Diffusion in Silicon in Inert and Oxidizing Ambient and Extrinsic Conditions
(66.30; S5.11)
(a) Vol. **100**, No. 2, p. 461—465 (1987).

B. Esser
*Random Walk on Ultrametric Spaces: A Time Dependent Bond Construction
(66.30)
(b) Vol. **141**, No. 1, p. K11—K14 (1987).

P. H. Dionisio and I. J. R. Baumvol
*The Influence of Tin Ion Implantation and Radiation-Enhanced Tin Diffusion on the Thermal Oxidation of a High-Carbon Steel
(66.30; 61.80; S1.1; S10)
(a) Vol. **101**, No. 1, p. 63—76 (1987).

G. Kluge
*A New Interpretation of the Photodoping Effect in Amorphous As- and Ge-Chalcogenides
(66.30; 68.48; S8)
(a) Vol. **101**, No. 1, p. 105—114 (1987).

M. Wienecke and M. Schenk
In-Situ High-Temperature Hall and Conductivity Measurements for Investigations of the Stability Region of PbTe
(61.70; 66.30; 72.80; S8)
(a) Vol. **101**, No. 1, p. 115—122 (1987).

G. Kluge, R. Klabes, A. Thomas, R. Grötzschel, and P. Süptitz
*High Energy Ion Beam Induced Ag Diffusion within Amorphous Ag:$GeSe_2$
(66.30; S8)
(a) Vol. **101**, No. 1, p. K9—K11 (1987).

P. Laborde, G. Villeneuve et J. M. Reau
*Etude par RMN de la mobilité de l'ion flour au sein de la solution solide $Pb_{1-x}Bi_xO_xF_{2-x}$
(NMR Study of the Mobility of Fluor Ions in $Pb_{1-x}Bi_xO_xF_{2-x}$ Solid Solutions)
(66.30; 76.60; S9)
(b) Vol. **141**, No. 2, p. 379—390 (1987).

A. Ya. Neiman, A. L. Podkorytov, and V. M. Zhukovskii
*Defect Structure and Transport Properties of $Me_6M_2O_{11}$ (Me:Sr, Ba; M:Nb, Ta)-Type Phases
(61.70 and 66.30; S10.15)
(a) Vol. **101**, No. 2, p. 371—380 (1987).

J. Kirchhof, P. Kleinert, W. Radloff, and E. Below
*Diffusion Processes in Lightguide Materials. The Diffusion of OH in Silica Glass at High Temperatures
(66.30; 64.75; S10.1)
(a) Vol. **101**, No. 2, p. 391—401 (1987).

R. Kirchheim and X. Y. Huang
*A Relationship between Prefactor and Activation Energy for Diffusion
(66.30)
(b) Vol. **144**, No. 1, p. 253—257 (1987).

G. Vogl
On the Motion of Solute Atoms in Cages and Loops in Metal Lattices Studied with Nuclear Methods
(61.70; 66.30; 76.80; S1; S1.3; S4)
(b) Vol. **144**, No. 1, p. 259—269 (1987).

N. Smetniansky-De Grande, E. J. Savino, and C. N. Tomé
*Stress Induced Anisotropic Diffusion of Intrinsic Point Defects towards Dislocations in H.C.P. Crystals
(61.70 and 66.30)
(b) Vol. **144**, No. 1, p. 271—286 (1987).

P. Vargas, H. Kronmüller, and M. C. Böhm
H_2 Dissociation in Nickel. A Cluster Model
(61.70; 66.30; S1.2)
(b) Vol. **144**, No. 1, p. 305—313 (1987).

J. Völkl, H. Wipf, B. J. Beaudry, and K. A. Gschneidner, Jr.
*Diffusion of H and D in Lutetium
(66.30; S1.4)
(b) Vol. **144**, No. 1, p. 315—327 (1987).

G. Neumann
*A Model for the Calculation of Monovacancy and Divacancy Contributions to the Impurity Diffusion in Noble Metals
(61.70 and 66.30; S1.2; S1.3)
(b) Vol. **144**, No. 1, p. 329—341 (1987).

G. O. Williams, V. A. Singh, and J. W. Corbett
Collision Cascades and Fractals
(61.80; 66.30)
(b) Vol. **144**, No. 1, p. K1—K5 (1987).

N. A. Stolwijk, J. Hölzl, W. Frank, J. Hauber, and H. Mehrer
*Decoration of Defects in Silicon with Gold, and Related Subjects
(66.30; 61.70; S5.11)
(a) Vol. **104**, No. 1, p. 225—245 (1987).

H. Mehrer
*Implantation of Radioisotopes at ISOLDE. A Novel Source Production Technique for Diffusion Studies in Solids
(66.30; 61.70; S1.2; S1.3; S5.11)
(a) Vol. **104**, No. 1, p. 247—261 (1987).

M.-P. Macht, A. Müller, V. Naundorf, and H. Wollenberger
Determination of Temperature-Dependent Sink Concentrations in Ion-Irradiated Nickel and Copper by Diffusion Measurements
(61.70; 61.80; 66.30; S1.2)
(a) Vol. **104**, No. 1, p. 287—295 (1987).

66.70. Nonelectronic thermal conduction and heat-pulse propagation in nonmetallic solids

(for thermal conduction in solid metals, see 72.15; for statistical mechanics of lattice vibrations see 63.70)

S. M. Wasim and R. Tovar B.
*On the Thermal Conductivity of $CuInSe_{2x}S_{2(1-x)}$
(66.70; S8.15)
(a) Vol. **99**, No. 2, p. K69—K73 (1987).

P. Enders
*Addendum to: Non-Fourierian Heat Conduction in Semiconductor Multilayer-Systems with Thin Layers — Asymptotic Behavior of Decay Constants
(66.70)
(b) Vol. **140**, No. 2, p. K97—K98 (1987).

G. D. Ivlev and V. L. Malevich
On Phase Transitions Stimulated in Amorphized Silicon by Nanosecond Pulsed Laser Annealing
(64.60; 66.70; S5.11)
(a) Vol. **103**, No. 2, p. K87—K91 (1987).

W. Mutscheller and M. Wagner
*The Influence of Magnetic Fields on the Thermal Conductivity of the Cooperative Vibronic System $TmVO_4$
(66.70; S11.1)
(b) Vol. **144**, No. 2, p. 507—518 (1987).

66.90. Other topics in nonelectronic properties

A. I. Sergienko
*Dynamics of Bjerrum Faults and Protonic Ice Conductivity
(66.10 and 66.90; 61.70; S10)
(b) Vol. **144**, No. 2, p. 471—475 (1987).

67. Quantum fluids and solids, liquid and solid helium

67.40. Boson degeneracy and superfluidity of helium-4

67.80. Solid helium and related quantum crystals

68. Surfaces and interfaces; thin films and whiskers
(for impact phenomena, see 79; for crystal growth, see 61.50)

68.20. Solid surface structures

A. A. KARPUSHIN, A. N. SOROKIN, I. YU. SEMENOVA, and M. TOMÁŠEK
*A Contribution to the Theoretical Interpretation of the UPS Spectra of Monoatomic Adsorption of Oxygen on the Si(111) Surface
(68.20 and 73.20; 79.60; S5.11)
(b) Vol. **139**, No. 1, p. 125—130 (1987).

M. PARLAPANSKI
*Thermionic Emission from Irregular Metal Surfaces
(68.20 and 79.40; S2)
(b) Vol. **139**, No. 1, p. K5—K8 (1987).

D. LOVRIĆ and B. GUMHALTER
Physisorption and Chemisorption Effects on Atom–Adsorbate van der Waals Potentials. Application to He and He Scattering from Adsorbed CO
(68.20 and 73.90; 79.20)
(b) Vol. **139**, No. 2, p. 423—432 (1987).

P. GODOWSKI and J. RUDNY
*Effect of Sulphur on the Surface Segregation in $Co_{50}Ni_{50}$ Alloy
(68.20; 79.20; S1.2)
(a) Vol. **99**, No. 2, p. 527—536 (1987).

A. TOUMI et C. LANDRON
*Influence de l'état de surface sur les calculs de correction en microanalyse X
(Influence of the State of the Surface on the Calculation of Corrections in Electron Probe Microanalysis)
(68.20)
(a) Vol. **100**, No. 2, p. 535—544 (1987).

I. E. TÜRE, A. W. BRINKMAN, G. J. RUSSELL, and J. WOODS
The Effects of Oxygen in Au–CdSe Schottky Diodes
(85; 68.20; 73.30; S8.12)
(a) Vol. **100**, No. 2, p. 681—691 (1987).

H. ENGLISCH and M. HERRMANN
*Adsorption at Random Chains
(68.20)
(b) Vol. **141**, No. 1, p. K41—K46 (1987).

I. A. ABROYAN, M. V. KRUGLOV, and I. K. SOLOMIN
Surface-Layer Swelling on Implantation of Hydrogen Ions into Silicon at Different Doping Concentrations
(61.80; 68.20; S5.11)
(a) Vol. **101**, No. 1, p. K5—K8 (1987).

J. ABELLÁN, R. CHICÓN, and M. ORTUÑO
*The Interpretation and Reconstruction of Images in Scanning Tunneling Microscopy
(68.20)
(a) Vol. **101**, No. 2, p. 463—468 (1987).

M. A. VASILIEV and S. D. GORODETSKII
Ferromagnetic Anomalies in the Surface of Single Crystals of the Fe–Ni System
(75.50; 68.20; S1.1; S1.2)
(a) Vol. **102**, No. 2, p. K121—K125 (1987).

P. RENNERT and F. SEIFERT
*Scattering Properties of Non-Spherical Muffin-Tin Potentials of Surface Atoms
(68.20 and 71.10; 73.20; S1.2)
(b) Vol. **144**, No. 2, p. 617—629 (1987).

68.25. Mechanical and acoustical properties of solid surfaces and interfaces

V. I. SOKOLOV and N. A. FEDOROVICH
*Mechanical Stresses on the Si–SiO_2 Interface
(68.25; S5.11; S10.1)
(a) Vol. **99**, No. 1, p. 151—158 (1987).

A. K. PAL and M. GANGULY
*Bleustein-Gulyaev Waves in Materials with Strain-Dependent Dielectric Constant
(68.25; 71.38)
(a) Vol. **103**, No. 1, p. 167—173 (1987).

68.30. Dynamics of solid surfaces and interface vibrations

V. G. PODOPRIGORA, I. A. REMIZOV, V. F. SHABANOV, and A. N. BOTVICH
*Surface Phonons in Molecular Crystals
(68.30; S12)
(b) Vol. **139**, No. 1, p. 109—115 (1987).

J. S. NKOMA
*Surface Phonon Polaritons in Semi-Infinite Semiconductor Superlattices
(68.30 and 71.36; S7.11; S7.12)
(b) Vol. **139**, No. 1, p. 117—123 (1987).

S. YA. VETROV and V. F. SHABANOV
*Dissipative Effects of the Interaction of Crystal Lattice Surface with Medium
(68.30)
(b) Vol. **140**, No. 1, p. 103—112 (1987).

PAN JIN-SHENG
Electron–Phonon Interaction in Dielectric Interface Systems
(77.55; 68.30; 71.38)
(b) Vol. **140**, No. 2, p. 421—436 (1987).

V. S. VIKHRENKO and M. I. KULAK
*Vibrational States of Surface Crystal Atoms
(68.30; S5.2)
(b) Vol. **140**, No. 2, p. 459—465 (1987).

G. WIATROWSKI
Stochastic Forces in Anharmonic Crystals and on Their Surfaces
(63.20; 68.30)
(b) Vol. **141**, No. 1, p. 85—96 (1987).

L. WENDLER and R. PECHSTEDT
Dynamical Screening, Collective Excitations, and Electron–Phonon Interaction in Heterostructures and Semiconductor Quantum Wells. Application to Double Heterostructures
(71.38 and 71.45; 68.30; 73.40; 77.30; 78.65; S7.12; S7.15)
(b) Vol. **141**, No. 1, p. 129—150 (1987).

L. WENDLER and R. HAUPT
Electron–Phonon Interaction in Semi-Infinite Semiconductor Superlattices
(71.38 and 71.45; 63.20; 68.30; 73.40; S7.1; S7.15)
(b) Vol. **141**, No. 2, p. 493—505 (1987).

R. HAUPT and L. WENDLER
Dispersion and Damping Properties of Plasmon Polaritons in Superlattice Structures. I. Infinite Superlattice
(71.36 and 71.45; 68.30)
(b) Vol. **142**, No. 1, p. 125—139 (1987).

M. BHATTACHARYA, S. MUKHOPADHYAY, and T. K. MITRA
Surface Optical Polaron in Magnetic Field
(71.38 and 73.20; 68.30)
(b) Vol. **142**, No. 1, p. 141—148 (1987).

R. HAUPT and L. WENDLER
Dispersion and Damping Properties of Plasmon Polaritons in Superlattice Structures. II. Semi-Infinite Superlattice
(71.36 and 71.45; 68.30; 78.65)
(b) Vol. **142**, No. 2, p. 423—435 (1987).

L. WENDLER and R. HAUPT
*An Improved Virtual Mode Theory of ATR Experiments on Surface Polaritions. Application to Long-Range Surface Plasmon-Polaritons in Asymmetric Layer Structures
(68.30 and 78.65; 71.36; 71.45; S1.3)
(b) Vol. **143**, No. 1, p. 131—148 (1987).

L. WENDLER and R. HAUPT
Electron–Phonon Interaction in Semiconductor Superlattices
(71.38; 63.20; 68.30; 73.40; S7.1; S7.15)
(b) Vol. **143**, No. 2, p. 487—510 (1987).

68.45. Solid–fluid interface processes

Yu. B. Bolkhovityanov and R. I. Bolkhovityanova
*A GaAs Surface after Its Contact with an Undersaturated Al–Ga–As Liquid. Observation after Anodic Oxidation
(68.45; S7.12)
(a) Vol. **99**, No. 1, p. 159—164 (1987).

T. D. Radzhabov, A. M. Nazarov, and U. U. Pulatov
*Surface Properties of Ion Alloyed Thin Dysprosium Films
(68.45; S1.4)
(a) Vol. **99**, No. 2, p. K83—K86 (1987).

V. S. Demidenko, A. Szász, and M. A. Aysawi
*On the Model Calculation of the Excitonic-Like States and Their Possible Role in Autocatalytic Processes. I. Application to the Formation of Autocatalytic Ni–P Amorphous Coatings
(68.45 and 68.55; 71.35; S1.2; S1.62)
(b) Vol. **140**, No. 1, p. 121—126 (1987).

W. Th. M. Buters, A. van Veen, and A. van den Beukel
Some Results on Helium Trapping in Undeformed and Cold Worked Single Crystalline Copper. A Comparison with Molybdenum and Nickel
(61.70; 68.45; 79.20; S1.2)
(a) Vol. **100**, No. 1, p. 87—94 (1987).

F. Lukeš
Temperature Dependence of Ellipsometric Angles of Silicon
(78.20; 68.45; S5.11)
(a) Vol. **102**, No. 2, p. 803—814 (1987).

B. Wolf and A. Zehe
*Reaction Probability of Molecular Deuterium with a Disordered InSb (110) Surface
(68.45; S7.13)
(a) Vol. **104**, No. 1, p. K33—K35 (1987).

W. Lorenz
*Chemical Bond Formation Processes on Semiconductor Interfaces at Electronic Equilibrium or Non-equilibrium
(68.45 and 73.40; 71.20; 73.30)
(b) Vol. **144**, No. 2, p. 585—599 (1987).

R. Hergt, H. Pfeiffer, P. Görnert, M. Wendt, B. Keszei, and J. Vandlik
Kinetic Segregation of Lead Impurities in Garnet LPE Films
(64.75; 68.45; 68.55; S11.2)
(a) Vol. **104**, No. 2, p. 769—776 (1987).

68.48. Solid–solid interfaces
(including bicrystals)
(for grain boundaries, see 61.70)

M. Ben Salem
General Conditions for Obtaining Coherent Walls under Strain between Elastic Domains
(64.80; 68.48)
(a) Vol. **99**, No. 2, p. 449—459 (1987).

M. Procop, H. Raidt, and B. Sandow
*Auger Electron and X-Ray Spectroscopic Analysis of Au–Ge Contacts to GaAs
(68.48; 73.40; 79.20; S1.3; S7.12)
(a) Vol. **99**, No. 2, p. 573—580 (1987).

B. Balland, A. Glachant, A. Ronda, and J. C. Dupuy
*High-Temperature Thermal Nitridation and Low-Temperature Electron-Beam-Enhanced Nitridation of SiO_2 Thin Films
(68.48 and 73.40; S5.11; S10.1)
(a) Vol. **100**, No. 1, p. 187—198 (1987).

R. Portius, D. Dietrich, A. Hässner, Th. Raschke, E. Wieser, and W. Wolke
*Formation of Stable $MoSi_2$/Poly-Si Films by Rapid Thermal Annealing
(68.48 and 73.60; S1.61; S5.11)
(a) Vol. **100**, No. 1, p. 199—206 (1987).

G. Kluge
A New Interpretation of the Photodoping Effect in Amorphous As- and Ge-Chalcogenides
(66.30; 68.48; S8)
(a) Vol. **101**, No. 1, p. 105—114 (1987).

Chan-Syun and Juh-Tzeng Lue
Palladium Silicide Formation on Amorphous Silicon by Thermal Annealing
(68.55; 68.48; S1.61)
(a) Vol. **101**, No. 1, p. 151—157 (1987).

N. Wagner, M. Horx, M. Jacobs, and M. Suchi
*XPS Investigations of TiO_2/RuO_2 Mixed Oxide Films on Titanium
(68.48 and 79.60; S10.15)
(a) Vol. **101**, No. 1, p. K23—K27 (1987).

L. Guoba, G. Zubauskas, J. Kameneckas, and A. Sakalas
*Study of Multilayer Structures on GaAs by Rutherford Backscattering with Carbon Ions
(68.48; 79.20; S7.12)
(a) Vol. **101**, No. 2, p. 445—450 (1987).

R. Ewertowski and W. Świątkowski
Annihilation of Positron Trapped at the Ag/Na_3AlF_6 Interface
(78.70; 68.48; S9.16)
(a) Vol. **102**, No. 2, p. 555—557 (1987).

P. K. Khanna, S. K. Bhatnagar, and M. L. Sisodia
*On the TCR and Electrical Conduction in a Thick-Film Segmented-Resistor Structure
(68.48 and 73.40; S10.1)
(a) Vol. **103**, No. 1, p. K17—K21 (1987).

M. Procop, B. Sandow, H. Raidt, and Li Do Son
The Importance of the Ni to Ge Ratio and of the Annealing Cycle for the Resistivity and Morphology of NiAuGe Ohmic Contacts to n-GaAs
(73.40 and 85; 68.48; 79.20; S7.12)
(a) Vol. **104**, No. 2, p. 903—916 (1987).

68.55. Thin film growth, structure, and epitaxy

H. Rübel, B. Schröder, W. Fuhs, J. Krauskopf, T. Rupp, and K. Bethge
*IR Spectroscopy and Structure of RF Magnetron Sputtered a-SiC:H Films
(68.55 and 73.60; 78.30; S6)
(b) Vol. **139**, No. 1, p. 131—143 (1987).

P. Schäfer and H. Niebsch
*Growth of (100) and (111) CdTe Films on (100) GaAs by Hot-Wall-Epitaxy
(68.55; S8.13)
(a) Vol. **99**, No. 1, p. K41—K44 (1987).

M. H. Islam and C. A. Hogarth
Some Optical and Structural Studies on SiO/SnO_2 Films
(78.65; 68.55; 71.25; S10)
(a) Vol. **99**, No. 2, p. 503—512 (1987).

H. Pfeiffer and R. Hergt
*Influence of Texture Formation on Properties of Films for Perpendicular Magnetic Recording
(68.55 and 75.70)
(a) Vol. **99**, No. 2, p. 549—558 (1987).

V. S. Demidenko, A. Szász, and M. A. Aysawi
*On the Model Calculation of the Excitonic-Like States and Their Possible Role in Autocatalytic Processes. I. Application to the Formation of Autocatalytic Ni–P Amorphous Coatings
(68.45 and 68.55; 71.35; S1.2; S1.62)
(b) Vol. **140**, No. 1, p. 121—126 (1987).

R. M. CÂNDEA, R. TURCU, G. BORODI, and I. BRATU
*Effects of Thermal Annealing in Air on VE, COD, and CAD PbSe Films
(68.55 and 73.60; 78.65; S8)
(a) Vol. **100**, No. 1, p. 149—155 (1987).

T. JUNG and M. SCHMIDT
*Plasma-Assisted Reactive Evaporation of Aluminium Nitride Films
(68.55 and 73.60; 79.20; S7.1)
(a) Vol. **100**, No. 1, p. 207—211 (1987).

J. ČERMÁK, L. KALIDOVA, and M. NEVŘIVA
*LPE Films of Magnetic Garnets with Bi Substitution
(68.55 and 75.70; S11.2)
(a) Vol. **100**, No. 1, p. 213—219 (1987).

A. SINGH and W. S. KHOKLE
*Sheet Resistance and Grain Size Study of Nickel Disilicide
(68.55 and 73.60; S1.61)
(a) Vol. **100**, No. 1, p. K25—K29 (1987).

R. P. SINGH and S. L. SINGH
Vacuum-Evaporated Zinc Phosphide Films and Their Characterization
(73.60; 68.55; 78.65; S7)
(a) Vol. **100**, No. 2, p. 493—500 (1987).

P. I. DIDENKO, A. A. EFREMOV, V. S. KHOMCHENKO, G. PH. ROMANOVA, and N. A. VLASENKO
*Comparative Study of SiO_x and SiO_x:TbF_3 Films
(68.55; 64.75; 78.65; S10.1)
(a) Vol. **100**, No. 2, p. 501—511 (1987).

J. GEORGE, B. PRADEEP, and K. S. JOSEPH
Preparation of Heat Mirrors Using Bismuth Oxide Films
(78.65; 68.55; S10.1)
(a) Vol. **100**, No. 2, p. 513—519 (1987).

B. KLOTH, M. V. LUPAL, A. N. PIKHTIN, C. E. RICHTER, R. RIES, R. STEGMANN, and M. TRAPP
*Determination of the Nitrogen Concentration in VPE-$GaAs_{1-x}P_x$
(68.55 and 78.65; 64.75; S7.15)
(a) Vol. **100**, No. 2, p. 545—552 (1987).

B. ROTHER, C. WEISSMANTEL, and G. LEONHARDT
*X-Ray Photoelectron Spectroscopy at Ion-Plated Boron Nitride
(68.55; 79.60; S7)
(a) Vol. **100**, No. 2, p. 553—557 (1987).

KH. A. MADY, H. A. MADY, S. Y. GIRGIS, and H. R. MOUSTAFA
*Effect of Deposition Rate on Vacuum-Deposited PbS Films
(68.55; S8)
(a) Vol. **100**, No. 2, p. K107—K109 (1987).

A. DRANDAROV and D. DOBREV
*Recrystallization Alterations in the Texture of Condensed Silver Layers
(68.55; S1.3)
(a) Vol. **102**, No. 2, p. K87—K90 (1987).

V. T. BICH, N. N. DINH, T. X. HOAI, N. H. HOANG, L. V. HONG, and V. D. MIEN
*Preparation of In_2O_3:Sn (ITO) Films by Electron Beam Deposition
(68.55; 73.60; 78.65; S10)
(a) Vol. **102**, No. 2, p. K91—K94 (1987).

S. V. VASILEV, N. N. GERASIMENKO, V. V. KALININ, and V. S. KULIKAUSKAS
*Redistribution of As during Platinum Silicide Formation
(66.30 and 68.55; S1.61)
(a) Vol. **103**, No. 1, p. 125—130 (1987).

K. YAMAMOTO and S. MARUYAMA
*A New Long Period Structure in Gold–Indium Films between 20 and 25 at% In
(68.55; S1.3; S4)
(a) Vol. **103**, No. 1, p. 153—160 (1987).

O. GOEDE, W. HEIMBRODT, V. WEINHOLD, E. SCHNÜRER, and H. G. EBERLE
Optical Study of Phase Transition from Tetrahedrally into Octahedrally Coordinated MnS
(78.65; 68.55; S1.63)
(b) Vol. **143**, No. 2, p. 511—518 (1987).

V. V. KOROBTSOV, V. G. LIFSHITS, A. V. ZOTOV, and V. G. SHENGUROV
*Solid Phase Epitaxy of Doped Silicon Films in Molecular Beam Epitaxy Systems
(68.55; S5.11)
(a) Vol. **103**, No. 2, p. 467—473 (1987).

N. SASI, C. BALASUBRAMANIAN, and S. K. NARAYANDASS
Structure, Dielectric, and AC Conduction Studies on Germanium Dioxide Thin Films
(68.55 and 73.60; 77.55; S10.1)
(a) Vol. **103**, No. 2, p. 475—480 (1987).

K. TENNAKONE, A. H. JAYATISSA, C. A. N. FERNANDO, S. WICKRAMANAYAKE, S. PUNCHIHEWA, L. K. WEERASENA, and W. D. R. PREMASIRI
Semiconducting and Photoelectrochemical Properties of n- and p-Type β-CuCNS
(73.60 and 78.65; 68.55; 72.40; S11)
(a) Vol. **103**, No. 2, p. 491—497 (1987).

T. STAPIŃSKI, E. JAPA, and J. ZUKROWSKI
*^{119}Sn Mössbauer Investigation of Cadmium–Tin–Oxide Thin Films
(68.55 and 76.80; S10.1)
(a) Vol. **103**, No. 2, p. K93—K98 (1987).

H. RATAJCZAK, I. GOŚCIAŃSKA, and A. SZLAFEREK
*The Influence of Isothermal Annealing on the Resistivity of Amorphous $Fe_{40}Ni_{40}B_{20}$ Thin Films
(61.40 and 68.55; 72.15; S1.1)
(a) Vol. **104**, No. 2, p. 731—741 (1987).

S. A. El-Sahhar, M. El-Shabasy, and H. M. Talaat
*Study of the Structure and Optical Properties of Thin Zinc Films
(68.55 and 78.65; S1.2)
(a) Vol. **104**, No. 2, p. 755—760 (1987).

R. Hergt, H. Pfeiffer, P. Görnert, M. Wendt, B. Keszei, and J. Vandlik
Kinetic Segregation of Lead Impurities in Garnet LPE Films
(64.75; 68.45; 68.55; S11.2)
(a) Vol. **104**, No. 2, p. 769—776 (1987).

68.90. Other topics in the structure and nonelectric properties of surfaces and thin films

V. V. Malashenko, V. L. Sobolev, and B. I. Khudik
Dynamical Deceleration of a Dislocation by Surface Defects
(61.70; 68.90)
(b) Vol. **144**, No. 2, p. 463—470 (1987).

70. Condensed matter: electronic structure; electrical, magnetic, and optical properties

71. Electronic states

(see also 63. Lattice dynamics, 73. Electronic structure and electrical properties of surfaces, interfaces, and thin films)

71.10. General theories and computational techniques

P. Ziesche and D. Lehmann
*Local Virial Relations and Pressure
(71.10)
(b) Vol. **139**, No. 2, p. 467—483 (1987).
Addendum
(b) Vol. **144**, No. 2, p. 911 (1987).

T. P. Eggarter and A. Troper
*A Simple Cluster Improvement to the Coherent Potential Approximation
(71.10; 71.25; S4; S12.1)
(b) Vol. **140**, No. 1, p. 127—133 (1987).

B. I. Craig and P. V. Smith
*Beyond the LUC-CNDO $k = 0$ Approximation
(71.10)
(b) Vol. **140**, No. 2, p. 491—500 (1987).

B. I. Craig and P. V. Smith
*A Semi-Empirical Approach to the Bulk Properties of Copper
(71.10 and 71.25; 62.20; 63.20; S1.2)
(b) Vol. **140**, No. 2, p. 501—508 (1987).

H. J. Schellnhuber and H. Urbschat
*Analyticity Breaking of Wave Functions and Fractal Phase Diagram for Simple Incommensurate Systems
(71.10; 75.20)
(b) Vol. **140**, No. 2, p. 509—519 (1987).

E. Kolley and W. Kolley
*Stochastic Quantization Approach to Disordered Fermions
(71.10; 72.70)
(b) Vol. **140**, No. 2, p. 527—533 (1987).

J. Monecke
*Material Constants of Composites. II. An Improvement over the Coherent Potential Approximation for Crystals with Site-Diagonal Disorder
(71.10; 63.50)
(b) Vol. **141**, No. 1, p. 173—176 (1987).

U. Lindner
To Applicability of Hamilton's Equations in Quantum Many-Body Problems
(75.10; 63.10; 71.10)
(b) Vol. **141**, No. 1, p. 263—268 (1987).

71.20. Electronic density of states determinations
(including energy states of liquid semiconductors)
(see also 65.40. Electronic heat capacity)

J. BULLOT and M. P. SCHMIDT
Physics of Amorphous Silicon–Carbon Alloys **(Review Article)**
(61.40; 71.20; 71.55; 72.40; 73.60; 78.20; 78.55; 78.60; S6)
(b) Vol. **143**, No. 2, p. 345—418 (1987).

H. SOLBRIG
*Recursive Scattered-Wave Approach to the Electronic Density of States of Topologically Disordered Systems
(71.20; S1.2)
(b) Vol. **139**, No. 1, p. 223—231 (1987).

J. BANHART, H. EBERT, and J. VOITLÄNDER
*Application of the Coherent Potential Approximation to Substitutional Ternary Alloys
(71.20; S1; S1.3)
(b) Vol. **139**, No. 1, p. K19—K23 (1987).

T. DRÜSEDAU, D. WEGENER, and R. BINDEMANN
*The Ordinary and Generalized Meyer-Neldel Rule in a-Si:H Calculated by the Statistical Fermi Level Shift
(71.20 and 72.20; S5.11)
(b) Vol. **140**, No. 1, p. K27—K31 (1987).

YU. A. FIRSOV, E. P. NAKHMEDOV, V. N. PRIGODIN, and W. WELLER
*Corrections from Electron–Electron Interaction to the One-Particle Density of States in Strongly Anisotropic Two-Dimensional Systems
(71.20; 73.40)
(b) Vol. **141**, No. 1, p. 111—118 (1987).

J. KIERUL, J. W. STASIAK, and M. KLUCZYNSKI
*Correlation and Influence of Phonons in Calculations of the Density of States in Thin Films
(71.20; 71.38; S1.2)
(b) Vol. **141**, No. 1, p. 119—128 (1987).

A. LATGÉ, S. S. MAKLER, and E. V. ANDA
*Electronic Properties of Amorphous Semiconductors. Lattice Topology Effects
(61.40 and 71.20; S5.11)
(b) Vol. **141**, No. 1, p. 163—171 (1987).

W. CORDTS, H. WIESNER, and J. MONECKE
*Pressure Dependence of the Electron Density in Silicon
(71.20; S5.11)
(b) Vol. **141**, No. 1, p. K53—K56 (1987).

J. ŠLECHTA
*On SCCF Treatment of the Hubbard Model of 1-D Alloys. Auger Densities
(71.20)
(b) Vol. **141**, No. 2, p. 457—466 (1987).

J. Šlechta
*On SCCF Treatment of the Nonorthogonality Effect in the Density of States of 1-D Binary Alloys
(71.20)
(b) Vol. **141**, No. 2, p. K101—K106 (1987).

W. Lorenz and T. Katterle
*Chemical Processes on Semiconductor Electrode Interfaces and Distribution of Charge Transfer on the Orbital Energy Scale
(71.20 and 73.20; S7.12)
(b) Vol. **142**, No. 1, p. 149—163 (1987).

H. Solbrig
*Effective-Atom Fluctuations and the Representative Effective Atom in Liquid Nickel. A Cluster Study within the Muffin-Tin Concept
(71.20; S1.2)
(b) Vol. **142**, No. 1, p. 203—209 (1987).

L. Dahmani, A. Brezini, M. Sebbani, and C. Deppolier
*Electronic Structure of Disordered Alloys. Effect of Off-Diagonal Disorder
(71.20)
(b) Vol. **142**, No. 2, p. 455—460 (1987).

I. Bâldea
*Phenomenological Treatment of the Departure from the Picture of Ideal One-Dimensional Conductor
(71.20; 71.38; 71.55; S12.1)
(b) Vol. **143**, No. 2, p. 557—567 (1987).

O. Foltin
Oblique Incidence Specular Reflectivity of Isotropic Semiconductors near Three-Dimensional Critical Points
(78.20; 71.20)
(b) Vol. **143**, No. 2, p. K151—K154 (1987).

H. Bross and R. Eder
*Self-Consistent MAPW Calculation with a Warped Muffin-Tin Potential.
I. The Electronic Structure of Al and Its Pressure Dependence
(71.20 and 71.25; 64.30; S4)
(b) Vol. **144**, No. 1, p. 175—193 (1987).

H.-G. Fritsche
*A Comparison of Electronic Properties of Clusters and Bulk of Copper and Palladium. Bandwidth, Density of States, Deformation Density
(71.20; 71.25; S1; S1.2)
(b) Vol. **144**, No. 1, p. 195—202 (1987).

U. Krey, H. Ostermeier, and J. Zweck
*Electronic Properties of Glassy Metals
(71.20; 72.15; S1.1; S1.2)
(b) Vol. **144**, No. 1, p. 203—212 (1987).

W. Lorenz
Chemical Bond Formation Processes on Semiconductor Interfaces at Electronic Equilibrium or Nonequilibrium
(68.45 and 73.40; 71.20; 73.30)
(b) Vol. **144**, No. 2, p. 585—599 (1987).

H. He
*Electronic and Vibrational Properties of One-Dimensional Quasi-Crystals
(63.20 and 71.20; S12.1)
(b) Vol. **144**, No. 2, p. 631—638 (1987).

H. Bross and R. Stryczek
*Self-Consistent MAPW Calculation with a Warped Muffin-Tin Potential. II. The Electronic Structure of Li and Its Pressure Dependence
(71.20 and 71.25; 64.30; S2)
(b) Vol. **144**, No. 2, p. 675—692 (1987).

J. Monecke
Material Constants of Composites. III. Development of the General Theory and Application to Alloys with Off-Diagonal Disorder
(71.10; 71.20; 71.25)
(b) Vol. **144**, No. 2, p. 693—700 (1987).

J. L. Jansons, V. J. Krumins, Z. A. Rachko, and J. A. Valbis
Luminescence Due to Radiative Transitions between Valence Band and Upper Core Band in Ionic Crystals (Crossluminescence)
(78.60; 71.20; S9.11; S9.12)
(b) Vol. **144**, No. 2, p. 835—844 (1987).

71.25. Nonlocalized single-particle electronic states

A. L. Ivanovskii, D. L. Novikov, and V. A. Gubanov
*Electronic Properties of Carbides, Nitrides, and Oxides of Subgroup VIa Transition Metals. Theory and Experiment **(Review Article)**
(71.25; 73.20; 78.70; 79.60; S1.61; S1.62; S10)
(b) Vol. **141**, No. 1, p. 9—33 (1987).

H. P. Bonde and S. B. Shrivastava
*Electron Momentum Density Distributions in Rhodium
(71.25; 78.70; S1; S1.3)
(b) Vol. **139**, No. 1, p. 155—160 (1987).

A. Masłowska
*The Burstein Absorption Edge Structure in Lead Chalcogenides
(71.25; 71.38; 78.20; S8)
(b) Vol. **139**, No. 1, p. 161—171 (1987).

B. V. Egorov, I. B. Egorova, and M. A. Krivoglaz
*The Electronic Heterogeneous States in Ferroelectric Semiconductors
(71.25 and 77.80; 65.50; 77.30)
(b) Vol. **139**, No. 1, p. 173—184 (1987).

M. Mondal and K. P. Ghatak
*Theoretical Analysis of the Effective Electron Mass in n-Channel Inversion Layers on Ternary Chalcopyrite Semiconductors
(71.25; S7.16)
(b) Vol. **139**, No. 1, p. 185—193 (1987).

M. H. ISLAM and C. A. HOGARTH
Some Optical and Structural Studies on SiO/SnO_2 Thin Films
(78.65; 68.55; 71.25; S10)
(a) Vol. **99**, No. 2, p. 503—512 (1987).

T. P. EGGARTER and A. TROPER
A Simple Cluster Improvement to the Coherent Potential Approximation
(71.10; 71.25; S4; S12.1)
(b) Vol. **140**, No. 1, p. 127—133 (1987).

J. MAŠEK and B. VELICKÝ
*Mn 3d States in Photolectron Spectra from $Cd_{1-x}Mn_xTe$
(71.25; 79.60; S8.15)
(b) Vol. **140**, No. 1, p. 135—140 (1987).

S. SZUSZKIEWICZ
*A Study of the Electronic Structure of $Fe_{1-x}Co_x$ Alloys by a Polarized Positron Annihilation
(71.25; 78.70; S1.1; S1.2)
(b) Vol. **140**, No. 1, p. 141—147 (1987).

P. H. E. MEIJER, P. PECHEUR, and G. TOUSSAINT
Electronic Structure of the Cd Vacancy in CdTe
(71.55; 61.70; 71.25; S8.13)
(b) Vol. **140**, No. 1, p. 155—162 (1987).

G. BULK and W. NOLTING
Many-Body Effects in Antiferromagnetic 4f-Systems
(75.10; 71.25)
(b) Vol. **140**, No. 1, p. 261—271 (1987).

A. STAHL and D. FRANK
Dynamics of Phonon-Assisted Optical Interband Processes in Semiconductors
(78.20; 71.25; 71.36)
(b) Vol. **140**, No. 1, p. 301—310 (1987).

L. K. MISRA and D. N. TRIPATHY
Estimation of Plasmon Dispersion-Coefficient and Evaluation of the Effective Mass of Electrons in Metals
(71.45; 71.25; S2; S4)
(b) Vol. **140**, No. 2, p. 477—490 (1987).

B. I. CRAIG and P. V. SMITH
*A Semi-Empirical Approach to the Bulk Properties of Copper
(71.10 and 71.25; 62.20; 63.20; S1.2)
(b) Vol. **140**, No. 2, p. 501—508 (1987).

H. STACHOWIAK
*Screening of a Proton Moving through the Electron Gas. II. Kinetic Properties of the Screening Cloud
(71.25; 78.70)
(b) Vol. **140**, No. 2, p. 521—526 (1987).

P. E. MIJNARENDS
*Electron and Positron States in Disordered Alloys
(71.25 and 71.60; 71.10; 78.70; S1.2; S1.3)
(a) Vol. **102**, No. 1, p. 31—46 (1987).

J. LÓPEZ GONDAR and M. DE DIOS LEYVA
*Electronic Structure and Symmetry Properties of the Envelope Functions in PbTe–$Pb_{1-x}Sn_xTe$ Superlattices
(71.25; S8.15)
(b) Vol. **142**, No. 2, p. 445—453 (1987).

C. F. VAN BRUGGEN
*Monovalent-Copper (I; $3d^{10}$, $S = 0$) Induced Oxygen-2p Valence Band Holes with Concomitant Attractive Interaction inside Low-Dimensional Oxygen Networks
(71.25 and 74.70; 75.20; S1.4; S10.15)
(b) Vol. **142**, No. 2, p. 537—550 (1987).

K. TAKARABE, H. KAWAMURA, and K. WAKAMURA
*Electronic Band Structure and Optical Anisotropy in InS. I. Experimental Study
(71.25 and 78.20; S8)
(b) Vol. **142**, No. 2, p. 605—615 (1987).

G. GOBSCH, D. SCHULZE, and G. PAASCH
*Inversion Layers in InSb Bicrystals with a Charged Grain Boundary of Finite Thickness
(71.25 and 73.40; S7.13)
(b) Vol. **142**, No. 2, p. K119—K124 (1987).

M. DE DIOS LEVYA, J. LÓPEZ GONDAR, and J. SABÍN DEL VALLE
*Nonparabolicity Parameter in the Conduction Band of GaAs–$Al_xGa_{1-x}As$ Quantum Wells
(71.25; S7.13; S7.15)
(b) Vol. **142**, No. 2, p. K151—K154 (1987).

C. AMMARGUELLAT, M. ESCORNE, A. MAUGER, E. BEAUREPAIRE, M. F. RAVET, G. KRILL, F. LAPIERRE, P. HAEN, and C. GODART
*Mixed Valence Properties of CeM_2Si_2 (M = Mn, Fe, Co, Ni, Cu)
(71.25; 72.15; 75.30; S1.61)
(b) Vol. **143**, No. 1, p. 159—166 (1987).

V. P. ZHUKOV, V. A. GUBANOV, O. JEPSEN, N. E. CHRISTENSEN, and O. K. ANDERSEN
*Calculated Electronic Structures and Chemical Bonding in Sc and Y Carbides
(71.25; S1.61)
(b) Vol. **143**, No. 1, p. 173—184 (1987).

S. MANHAS, K. C. KHULBE, D. J. S. BECKETT, G. LAMARCHE, and J. C. WOOLLEY
Lattice Parameters, Energy Gaps, and Magnetic Properties of the $Cd_xHg_yMn_zTe$ Alloy System
(75.30; 61.60; 71.25; S8.16)
(b) Vol. **143**, No. 1, p. 267—274 (1987).

J. E. LEWIS
Optical Properties of the Alloy System SnTe–GeTe from Reflectance Measurements
(78.20; 71.25; 78.30; S8)
(b) Vol. **143**, No. 1, p. 307—315 (1987).

A. STESMANS
Interaction between Localized and Delocalized States Indicated from Temperature-Dependent ESR Study of Submetallic P-Doped Si
(76.30; 71.25; S5.11)
(b) Vol. **143**, No. 2, p. 733—740 (1987).

H. BOUDRIOT, K. DEUS, R. GRÜNDLER, H. A. SCHNEIDER, and T. STÖCKERT
Optical Properties of $ZnSiAs_2$ Single Crystalline Thin Films from Electron Energy Loss Measurements
(78.20 and 79.20; 71.25; S7.16)
(b) Vol. **143**, No. 2, p. K125—K130 (1987).

D. E. POPOV, E. I. ROZUM, S. E. SAVUSHKIN, S. R. UGLOV, and S. A. VOROBIEV
*Optical Radiation of Electrons and Positrons at Quasi-Molecular Channeling in Crystals
(71.25 and 78.70)
(b) Vol. **143**, No. 2, p. K141—K145 (1987).

H. BROSS and R. EDER
*Self-Consistent MAPW Calculation with a Warped Muffin-Tin Potential. I. The Electronic Structure of Al and Its Pressure Dependence
(71.20 and 71.25; 64.30; S4)
(b) Vol. **144**, No. 1, p. 175—193 (1987).

H.-G. FRITSCHE
A Comparison of Electronic Properties of Clusters and Bulk of Copper and Palladium. Bandwidth, Density of States, Deformation Density
(71.20; 71.25; S1; S1.2)
(b) Vol. **144**, No. 1, p. 195—202 (1987).

K. HASHIZUME
*Band Structure of Solid Argon within Relativistic APW Approximation
(71.25; S5.2)
(b) Vol. **144**, No. 1, p. K23—K28 (1987).

M. ZAŁUŻNY and A. ŁUKASIK
*The Influence of the Size Quantization on the Fermi Energy and the Effective Mass in Bi Films
(71.25; S4)
(b) Vol. **144**, No. 2, p. 575—584 (1987).

R. PÉREZ-ALVAREZ, J. L. PARRA-SANTIAGO, and P. PAJÓN-SUÁREZ
*Cylindrical Quantum Well with Position Dependent Mass
(71.25; S7.12; S7.15)
(b) Vol. **144**, No. 2, p. 639—644 (1987).

S. M. RYABCHENKO, YU. G. SEMENOV, and O. V. TERLETSKII
*Exchange-Scattering Effects on Band Energies in Magnetically Mixed Semiconductors
(71.25; 71.35; 75.30; S8.15)
(b) Vol. **144**, No. 2, p. 661—673 (1987).

H. BROSS and R. STRYCZEK
*Self-Consistent MAPW Calculation with a Warped Muffin-Tin Potential. II. The Electronic Structure of Li and Its Pressure Dependence
(71.20 and 71.25; 64.30; S2)
(b) Vol. **144**, No. 2, p. 675—692 (1987).

71.30. Metal–insulator transitions

71.35. Excitons and related phenomena
(including electron–hole dròps)

G. MANZKE, K. HENNEBERGER, and V. MAY
*Many-Exciton Theory for Multiple Quantum-Well Structures
(71.35; 78.20; S7.12; S7.15)
(b) Vol. **139**, No. 1, p. 233—239 (1987).

R. KINK, K. KALDER, A. LÔHMUS, and H. NIEDRAIS
Effect of Pressure on the Luminescence of Trapped Excitons in Xe Crystals
(78.60; 71.35; 71.50; S5.2)
(b) Vol. **139**, No. 1, p. 321—327 (1987).

G. SEITFUDEM
Time-Resolved Investigation of the Ne_2^* Luminescence
(78.60; 71.35; S5.2)
(b) Vol. **139**, No. 1, p. 347—356 (1987).

M. SCHMIDT and G. RÖPKE
*The Pair Correlation Function and Mott Transition in Two-Component Fermi Systems
(71.35; 64.10)
(b) Vol. **139**, No. 2, p. 441—455 (1987).

T. KESSLER, R. MARKUS, H. NAHME, and N. SCHWENTNER
Nonlinear Excitation of Pure and Doped Rare Gas Crystals
(78.20; 71.35; 78.55; S5.2)
(b) Vol. **139**, No. 2, p. 619—625 (1987).

M. M. MERKIN and A. A. TSERTSVADZE
Exoemission Theory: The Exciton Model
(79.75; 61.70; 71.35; S9.11)
(b) Vol. **140**, No. 1, p. 73—82 (1987).

V. S. DEMIDENKO, A. SZÁSZ, and M. A. AYSAWI
On the Model Calculation of the Excitonic-Like States and Their Possible Role in Autocatalytic Processes. I. Application to the Formation of Autocatalytic Ni–P Amorphous Coatings
(68.45 and 68.55; 71.35; S1.2; S1.62)
(b) Vol. **140**, No. 1, p. 121—126 (1987).

G. CZAJKOWSKI and P. SCHILLAK
Interference Spectra of Semiconducting Slabs at Excitonic Resonances Studied by Stahl's Coherent Wave Approach to Band Gap Electrodynamics
(78.20; 71.35; S8.11; S8.12)
(b) Vol. **140**, No. 1, p. 273—278 (1987).

F. LISSE, J. KÖHLER, H. PUFAHL, and D. SCHMID
Sensitized Luminescence of Pure and Doped $NaNO_2$ Single Crystals. I. Stationary Spectra
(78.55; 71.35; 71.55; S11.1)
(b) Vol. **140**, No. 2, p. 605—619 (1987).

P. E. Simmonds and R. Sooryakumar
*Stress Effects on Acceptor Bound Excitons in GaAs
(71.35 and 71.55; 78.55; S7.12)
(b) Vol. **142**, No. 2, p. K137—K141 (1987).

A. N. Trukhin
Temperature Dependence of Luminescence Decay Kinetics of Self-Trapped Excitons, Germanium and Aluminium Centres in Crystalline Quartz
(78.55; 71.35; S10.1)
(b) Vol. **143**, No. 1, p. K83—K88 (1987).

U. Woggon, F. Henneberger, and H. Weinert
*On the Bleaching of the A-Bound Exciton Absorption in GaP: N. An Example for an Electron–Hole Liquid Induced Optical Nonlinearity
(71.35 and 78.55; S7.11)
(b) Vol. **143**, No. 2, p. 611—623 (1987).

K. M. Paraskevopoulos, C. Julien, and M. Balkanski
Fine Structure in the Free Exciton Region of the Emission Spectrum of γ-InSe
(78.60; 71.35; S8)
(b) Vol. **143**, No. 2, p. 741—748 (1987).

J. L. Subtil, P. Laporte, R. Reininger, and V. Saile
VUV Optical Constants of Liquid and Solid Xenon at the Triple Point
(78.20; 71.35; S5.2)
(b) Vol. **143**, No. 2, p. 783—793 (1987).

V. A. Aliev, E. F. Bagirzade, N. Z. Gasanov, and G. D. Guseinov
Behaviour of the Urbach Edge at Phase Transitions in $TlGaSe_2$ Single Crystals
(78.40; 64.70; 71.35; S8.16)
(b) Vol. **143**, No. 2, p. K155—K160 (1987).

Ch. Khamsasy, Nguyen Van Hieu, and Nguyen Ai Viet
*Three-Branch Excitonic Polaritons in Direct Band Gap Cubic Semiconductors
(71.35 and 71.36; S7.1)
(b) Vol. **144**, No. 2, p. 645—652 (1987).

S. M. Ryabchenko, Yu. G. Semenov, and O. V. Terletskii
Exchange-Scattering Effects on Band Energies in Magnetically Mixed Semiconductor
(71.25; 71.35; 75.30; S8.15)
(b) Vol. **144**, No. 2, p. 661—673 (1987).

A. Haufe
*Nitrogen Bound Electrons, Excitons, and Multiexcitons in GaP
(71.35 and 71.55; S7.11)
(b) Vol. **144**, No. 2, p. 733—738 (1987).

71.36. Polaritons
(including photon–phonon and photon–magnon interactions)

71.38. Polarons and electron–phonon interactions
(see also 63.20. phonon-electron interactions in lattices)

A. MASŁOWSKA
The Burstein Absorption Edge Structure in Lead Chalcogenídes
(71.25; 71.38; 78.20; S8)
(b) Vol. **139**, No. 1, p. 161—171 (1987).

V. ČÁPEK
*Relaxation Superoperator Approach to the Generalized Master Equations in the Electron–Phonon Systems
(71.38)
(b) Vol. **139**, No. 2, p. K117—K121 (1987).

R. BOYN and H. ZIMMERMANN
"Dissolution" of 4f–4f Transitions in Phonon Sidebands (ZnS:Tm^{3+})
(71.55 and 78.55; 71.38; S8.11)
(b) Vol. **140**, No. 1, p. 163—172 (1987).

PAN JIN-SHENG
Electron–Phonon Interaction in Dielectric Interface Systems
(77.55; 68.30; 71.38)
(b) Vol. **140**, No. 2, p. 421—436 (1987).

V. ŽELEZNÝ, J. PETZELT, and R. SWIETLIK
Infrared Properties of $TeEA(TCNQ)_2$
(78.30; 63.20; 71.38; S12.1)
(b) Vol. **140**, No. 2, p. 595—604 (1987).

J. KIERUL, J. W. STASIAK, and M. KLUCZYNSKI
Correlation and Influnce of Phonons in Calculations of the Density of States in Thin Films
(71.20; 71.38; S1.2)
(b) Vol. **141**, No. 1, p. 119—128 (1987).

L. WENDLER and R. PECHSTEDT
*Dynamical Screening, Collective Excitations, and Electron–Phonon Interaction in Heterostructures and Semiconductor Quantum Wells. Application to Double Heterostructures
(71.38 and 71.45; 68.30; 73.40; 77.30; 78.65; S7.12; S7.15)
(b) Vol. **141**, No. 1, p. 129—150 (1987).

M. ASCHE und O. G. SARBEI
*Nonequilibrium Dynamics of Carriers and Phonons in GaAs after High Excitation by Short Time Laser Pulses
(71.38; 72.30; S7.12)
(b) Vol. **141**, No. 2, p. 487—491 (1987).

L. WENDLER and R. HAUPT
*Electron–Phonon Interaction in Semi-Infinite Semiconductor Superlattices
(71.38 and 71.45; 63.20; 68.30; 73.40; S7.1; S7.15)
(b) Vol. **141**, No. 2, p. 493—505 (1987).

M. ONODA and H. NAGASAWA
*Phase Transition and Bipolaron Formation in Quasi-One-Dimensional Conductor β-$Ag_xV_2O_5$
(64.60 and 71.38; 76.30; 76.70; S11)
(b) Vol. **141**, No. 2, p. 507—516 (1987).

71.45. Collective effects

I. A. Ovidko and A. E. Romanov
Topological Excitations (Defects, Solitons, Textures, Frustrations) in Condensed Media **(Review Article)**
(61.70; 64.60; 71.45; 75.30)
(a) Vol. **104**, No. 1, p. 13—45 (1987).

Tran Thoai Duy Bao
*Plasmon Excitation in the Interaction of Fast Electrons with Metal Surfaces
(71.45 and 73.20; 61.80; S4)
(b) Vol. **139**, No. 1, p. 145—154 (1987).

R. Gonczarek
*The Types of Quasiparticle Interactions in the Landau Theory
(71.45)
(b) Vol. **139**, No. 1, p. K15—K18 (1987).

L. A. Hussain and O. Akinlade
*A Model Potential and Its Applications
(71.45 and 72.15; 75.20)
(b) Vol. **139**, No. 2, p. K123—K128 (1987).

R. Gründler, T. Stöckert, H. Boudriot, K. Deus, and H. A. Schneider
*Volume Plasmon Dispersion of Polycrystalline Films of the Ternary Semiconductors $ZnSnAs_2$ and $ZnSiAs_2$
(71.45; S7.16)
(b) Vol. **140**, No. 1, p. K19—K22 (1987).

P. M. Karageorgy-Alkalaev and A. Yu. Leiderman
*The Synergetic Phenomena in Photoexcited Semiconductors
(71.45; 71.55; 72.40)
(a) Vol. **100**, No. 1, p. 221—231 (1987).

Tran Thoai Duy Bao
Image Force and Stopping Power of a Fast Moving Ion near a Metal Surface
(61.80 and 73.20; 71.45; S4)
(b) Vol. **140**, No. 2, p. 449—457 (1987).

Z. Ivić, M. Satarić, Z. Shemsedini, and R. Žakula
*The Role of Supersonic Davydov Solitons for Energy Exchange between Two Joined Molecular Chains
(71.45; S12.1)
(b) Vol. **140**, No. 2, p. 467—476 (1987).
Erratum
(b) Vol. **143**, No. 2, p. 806 (1987).

T. KOSTYRKO
*Collective Excitations in the Extended Hubbard Model with Intrasite Attraction
(71.45)
(b) Vol. **143**, No. 1, p. 149—158 (1987).

S. L. DRECHSLER, E. HEINER, and J. MALEK
Selfconsistent Discrete Equations for Nonlinear Defects in Peierls Systems with Broken Electron-Hole Symmetry
(71.50; 61.70; 71.45; S12.1)
(b) Vol. **143**, No. 1, p. 185—193 (1987).

C. O. RODRÍGUEZ, R. A. CASALI, E. L. PELTZER Y BLANCÁ, and O. M. CAPPANNINI
*First Principles Prediction of Structural Properties and Pressure Dependence of the Charge Density and Energy Gaps in Diamond
(61.50 and 71.45; 71.25; S5)
(b) Vol. **143**, No. 2, p. 539—548 (1987).

V. N. KADANTSEV, L. N. LUPICHOV, and A. V. SAVIN
*Intramolecular Excitation Dynamics in a Thermalized Chain. I. Formation of Autolocalized States in a Cyclic Chain
(71.45)
(b) Vol. **143**, No. 2, p. 569—579 (1987).

S. H. SINHA and A. R. CHETAL
*Correlation between Chemical Shift, Plasmon Energy, and Partial Charge
(71.45; S1.2)
(b) Vol. **143**, No. 2, p. 595—600 (1987).

L. S. BRIZHIK and A. S. DAVYDOV
Soliton Mechanism of Superconductivity in Organic Quasi-One-Dimensional Crystals
(74.20; 71.45; S12.1)
(b) Vol. **143**, No. 2, p. 689—698 (1987).

H. PUFF and H. W. STREITWOLF
Optical Absorption of the Soliton Lattice
(78.20; 71.45; S12)
(b) Vol. **143**, No. 2, p. 765—775 (1987).

J. FRIEDEL
On Electron Correlations of σ and π Bonding in Tetravalent Clusters and Solids
(71.10; 71.45)
(b) Vol. **144**, No. 1, p. 165—174 (1987).

71.50. Localized single-particle electronic states
(excluding impurities)

R. Kink, K. Kalder, A. Lõhmus, and H. Niedrais
Effect of Pressure on the Luminescence of Trapped Excitons in Xe Crystals
(78.60; 71.35; 71.50; S5.2)
(b) Vol. **139**, No. 1, p. 321—327 (1987).

J. C. Parlebas, K. Chetouane, and J. Hugel
*4f Optical Absorption Model in Rare-Earth Semiconducting Compounds
(71.50 and 78.40; S1.63)
(b) Vol. **142**, No. 1, p. 311—316 (1987).

R. A. Evarestov and V. P. Smirnov
*Symmetry of Localized Crystalline Orbitals and Double-Valued Band Representation
(71.50; S5.12)
(b) Vol. **142**, No. 2, p. 493—499 (1987).

S. L. Drechsler, E. Heiner, and J. Malek
*Selfconsistent Discrete Equations for Nonlinear Defects in Peierls Systems with Broken Electron–Hole Symmetry
(71.50; 61.70; 71.45; S12.1)
(b) Vol. **143**, No. 1, p. 185—193 (1987).

G. Seifert
*On the Electronic Shell Structure of Copper Clusters
(71.50; S1.2)
(b) Vol. **143**, No. 1, p. K37—K42 (1987).

A. A. Berezin
*Localized Electron Levels in Isotopically Ordered Crystals (Isotopic Quantum Wells)
(71.50; S10.1)
(b) Vol. **144**, No. 2, p. 727—732 (1987).

71.55. Impurity and defect levels

J. BULLOT and M. P. SCHMIDT
Physics of Amorphous Silicon–Carbon Alloys **(Review Article)**
(61.40; 71.20; 71.55; 72.40; 73.60; 78.20; 78.55; 78.60; S6)
(b) Vol. **143**, No. 2, p. 345—418 (1987).

H.-J. FÖCKEL
*Stationary and Resonance States at the Core of a Screw Dislocation and Their Contributions to Electrical Resistivity
(61.70 and 71.55; 72.15; S2)
(b) Vol. **139**, No. 1, p. 59—66 (1987).

J. LAKSHMANA RAO
*Optical Absorption Spectrum of Cr^{3+} Ions Doped in NH_4Cl Single Crystals
(71.55; 78.50; S9)
(b) Vol. **139**, No. 1, p. 241—244 (1987).

P. P. BODIUL, V. F. GARABAZHIU, and D. V. GITSU
*The Impurity States in Bismuth
(71.55; 71.70; S4)
(b) Vol. **139**, No. 1, p. 245—256 (1987).

D. L. NOVIKOV, A. L. IVANOVSKII, and V. A. GUBANOV
*The Influence of Carbon Vacancies and Local Atomic Displacements on the Electronic Structure of Niobium Carbides
(71.25 and 71.55; S1.61)
(b) Vol. **139**, No. 1, p. 257—265 (1987).

V. K. BASHENOV, D. L. KARDASHEV, and D. I. MARVAKOV
*Orbital Removal Method for the Neutral Vacancy in Semiconductors
(71.55; S5)
(b) Vol. **139**, No. 1, p. K31—K35 (1987).

H. LEMKE
*Dotierungseigenschaften von Nickel in Silizium
(Doping Properties of Nickel in Silicon)
(71.55; S5.11)
(a) Vol. **99**, No. 1, p. 205—213 (1987).

O. BREITENSTEIN and L. J. GILING
*Scanning-DLTS Investigations on Semi-Insulating GaAs: Cr, In Containing "Streamers"
(71.55; S7.12)
(a) Vol. **99**, No. 1, p. 215—223 (1987).

J. PASTRŇÁK, F. KAREL, J. OSWALD, W. ULRICI, and N. M. KOLCHANOVA
*The Study of Impurity Levels in LEC Semi-Insulating GaAs: Cr Samples by Two Source Photoconductivity Measurements
(71.55; 72.40; 78.55; S7.12)
(a) Vol. **99**, No. 1, p. 225—235 (1987).

I. THURZO and E. PINČÍK
Attempts to Find a Correlation between the GaAs Plasma Oxide Growth Conditions and the Charge DLTS Response of Interface States
(73.40; 71.55; S7.12)
(a) Vol. **99**, No. 2, p. K91—K95 (1987).

K. SCHMALZ, K. TITTELBACH-HELMRICH, and H. RICHTER
*On DLTS-Measurements of the Barrier-Limited Capture Process
(71.55 and 73.30; S5.11)
(a) Vol. **99**, No. 2, p. K97—K102 (1987).

P. H. E. MEIJER, P. PECHEUR, and G. TOUSSAINT
*Electronic Structure of the Cd Vacancy in CdTe
(71.55; 61.70; 71.25; S8.13)
(b) Vol. **140**, No. 1, p. 155—162 (1987).

R. BOYN and H. ZIMMERMANN
*"Dissolution" of 4f–4f Transitions in Phonon Sidebands (ZnS:Tm^{3+})
(71.55 and 78.55; 71.38; S8.11)
(b) Vol. **140**, No. 1, p. 163—172 (1987).

K. SCHMALZ, F.-G. KIRSCHT, S. NIESE, H. RICHTER, M. KITTLER, W. SEIFERT, I. BABANSKAYA, H. KLOSE, K. TITTELBACH-HELMRICH, and J. SCHÖNEICH
On the Intrinsic Gettering in Fe-Contaminated Cz–Si
(61.70; 71.55; S5.11)
(a) Vol. **100**, No. 1, p. 69—85 (1987).

P. M. KARAGEORGY-ALKALAEV and A. YU. LEIDERMAN
The Synergetic Phenomena in Photoexcited Semiconductors
(71.45; 71.55; 72.40)
(a) Vol. **100**, No. 1, p. 221—231 (1987).

B. RAY, P. A. MONSEF MIRZAI, and J. W. BRIGHTWELL
*Fluorescence and ESR in CaS:Mn^{2+}
(71.55; 76.30; 78.55; S8)
(a) Vol. **100**, No. 1, p. 233—237 (1987).

N. BALASUBRAMANYAM and V. KUMAR
*On the Quenched-In Defects in n-Type Silicon
(71.55; 62.50; S5.11)
(a) Vol. **100**, No. 1, p. 239—244 (1987).

J. KRYNICKI, J. C. MULLER, P. SIFFERT, I. BRYLOWSKA, and K. PAPROCKI
*Study of Defects in Silicon after Low Energy H^+ Implantation by DLTS Measurements
(61.70 and 71.55; 72.80; S5.11)
(a) Vol. **100**, No. 1, p. 245—249 (1987).

W. Ulrici, L. Eaves, K. Friedland, and D. P. Halliday
*Optical Studies of Vanadium in Gallium Phosphide
(71.55; 72.80; 78.50; 78.55; S7.11)
(b) Vol. **141**, No. 1, p. 191—202 (1987).

T. K. Pramanik and D. P. Bhattacharya
Impurity Breakdown in Germanium at Low Lattice Temperatures
(72.20; 71.55; S5.12)
(b) Vol. **141**, No. 1, p. 221—228 (1987).

S. V. Gorbunov, A. V. Kruzhalov, and M. J. Springis
Optical Properties of the F-Centre in Beryllium Oxide
(78.50 and 78.55; 61.70; 71.55; S10.1)
(b) Vol. **141**, No. 1, p. 293—301 (1987).

H. Pufahl, J. Köhler, Th. Schmidt, D. Schmid, and V. M. Kenkre
Sensitized Luminescence of Pure and Doped $NaNO_2$ Single Crystals. II. Time-Resolved Experiments and Preliminary Analysis
(78.55; 71.35; 71.55; S11.1)
(b) Vol. **141**, No. 1, p. 303—315 (1987).

H. Lemke
*Störstellenreaktionen in den Randschichtbereichen von Si-Kristallen mit Zn- und Pt-Dotierung
(Impurity Reactions in the Surface Regions of Zn and Pt Doped Si Crystals)
(61.70 and 71.55; S5.11)
(a) Vol. **101**, No. 1, p. 193—203 (1987).

M. Matos
UHF MC INDO Calculations of the Interstitial Hydrogen in Alkali Halides
(61.70; 71.55; S9.11)
(b) Vol. **141**, No. 2, p. 399—408 (1987).

F. J. Lohmeier
*Photostimulated Reorientations of PO_2^- Centers in Alkali Halides
(71.55; 78.55; S9.11)
(b) Vol. **141**, No. 2, p. 517—521 (1987).

W. Kühn, R. Strehlow, and M. Hanke
*Self-Consistent Tight-Binding Investigation of Chemical Trends for Native Defects in III–V Semiconductors
(71.55; S7.1)
(b) Vol. **141**, No. 2, p. 541—557 (1987).

F. Dujardin, B. Stébé, and G. Munschy
*Neutral Bound Excitons in a High Magnetic Field
(71.35 and 71.55)
(b) Vol. **141**, No. 2, p. 559—566 (1987).

V. V. Aristov, I. E. Bondarenko, J. Heydenreich, I. I. Khodos, I. I. Snighireva, P. Werner, E. B. Yakimov, and N. A. Yarykin
Electrical Properties and Defect Structure of Plastically Deformed Silicon Crystals Doped with Gold
(61.70; 71.55; S5.11)
(a) Vol. **102**, No. 2, p. 687—695 (1987).

V. K. Jain and M. S. Jahan
*Thermal Activation Energies and Peak Temperatures in Thermoluminescence of LiF(Mg, Ti) and CaF_2: Mn at Low Temperatures
(71.55 and 78.60; S9.11; S9.12)
(a) Vol. **102**, No. 2, p. 725—731 (1987).

W. Ulrici, J. Kreissl, A. Vasson, A. M. Vasson, and M. En-Naqadi
*Photoinduced Recharging Phenomena and Vanadium-Related Centres in GaAs: V
(71.55; 76.30; 78.50; S7.12)
(b) Vol. **143**, No. 1, p. 195—206 (1987).

J. Kreissl, W. Gehlhoff, and W. Ulrici
*Electron Paramagnetic Resonance of Titanium in GaP
(71.55 and 76.30; S7.11)
(b) Vol. **143**, No. 1, p. 207—215 (1987).

J. Pastrňák, F. Karel, and W. Ulrici
Two-Light-Source Extrinsic Photoconductivity of GaAs: V
(72.40; 71.55; S7.12)
(a) Vol. **103**, No. 1, p. 205—212 (1987).

I. Bâldea
Phenomenological Treatment of the Departure from the Picture of Ideal One-Dimensional Conductor
(71.20; 71.38; 71.55; S12.1)
(b) Vol. **143**, No. 2, p. 557—567 (1987).

Z. Badirkhan and J. Khalifeh
*Charge Transfer in Nickel–Hydrogen System
(71.55; S1.2)
(b) Vol. **143**, No. 2, p. 637—645 (1987).

M. Hanke, D. Hennig, and A. Kaschte
*A Tight-Binding Description of Isovalent Impurity Clusters. Application to Te Impurities in II–VI Compound Semiconductors
(71.55; S8.11; S8.12; S8.13)
(b) Vol. **143**, No. 2, p. 655—661 (1987).

J. L. Rao, R. M. Krishna, and S. V. J. Lakshman
*Nearest-Neighbour Point Ion Approximation Calculations of Energy Levels for Mn^{2+} Ions Doped in MgF_2 Single Crystals
(71.55; 78.50; S9.12)
(b) Vol. **143**, No. 2, p. K99—K101 (1987).

L. N. KANTOROVICH
*Multipole Theory of the Polarization of Solids by Point Defects. IV. Electronic and Spatial Structure of Single Electronic and Hole Centers and Their Pairs in LiF, KCl Crystals
(71.55; 77.30; S9.11)
(b) Vol. **144**, No. 2, p. 719—726 (1987).

A. HAUFE
*Nitrogen Bound Electrons, Excitons, and Multiexcitons in GaP
(71.35 and 71.55; S7.11)
(b) Vol. **144**, No. 2, p. 733—738 (1987).

YU. E. PERLIN, V. M. POLYANOVSKII, and O. V. ALIFANOV
*Optical Transitions of Jahn-Teller Local Centres
(71.55 and 78.50)
(b) Vol. **144**, No. 2, p. 817—824 (1987).

A. B. GERASIMOV, Z. G. GOGUA, D. A. KIMERIDZE, and V. N. SVANIDZE
*The Role of the Central Cell in the Problem of Carrier Capture by Attractive Centres in Ge and Si
(71.55; S5.11; S5.12)
(b) Vol. **144**, No. 2, p. K99—K104 (1987).

E. RADZHABOV
Vacuum Ultraviolet Absorption and Dissociation of H_a^- and OH^- Centers in NaF
(78.50; 61.70; 71.55; S9.11)
(b) Vol. **144**, No. 2, p. K151—K155 (1987).

A. N. ISOTOV and E. A. STEINMAN
*Reconstruction of Optical Dislocation Centres under the Action of Shear Stresses
(61.70 and 71.55; 78.55; S5.11)
(a) Vol. **104**, No. 2, p. 777—784 (1987).

D. LASZIG and P. HAASEN
Internal Friction of GaAs with EL2 Defects
(62.40; 71.55; S7.12)
(a) Vol. **104**, No. 2, p. K105—K109 (1987).

L. DÓZSA, V. ŠMÍD, J. J. MAREŠ, P. HUBÍK, and J. KRIŠTOFIK
Effect of the Interface of n^+-Type Substrate and n-Type Epitaxial Layer on the Junction Capacitance Spectroscopy
(73.40; 71.55; S7.12)
(a) Vol. **104**, No. 2, p. K117—K120 (1987).

P. S. GLADKOV, E. N. MONOVA, and K. B. OZANYAN
*LPE Growth and Characterization of GaAs:V
(71.55; 72.80; 78.55; S7.12)
(a) Vol. **104**, No. 2, p. K121—K126 (1987).

71.60. Positron states
(see also 78.70. Positron annihilation)

M. J. Puska
*Theoretical Aspects of Positrons in Imperfect Solids **(Review Article)**
(71.60)
(a) Vol. **102**, No. 1, p. 11—29 (1987).

P. E. Mijnarends
*Electron and Positron States in Disordered Alloys
(71.25 and 71.60; 71.10; 78.70; S1.2; S1.3)
(a) Vol. **102**, No. 1, p. 31—46 (1987).

H.-E. Schaefer
Investigation of Thermal Equilibrium Vacancies in Metals by Positron Annihilation
(61.70; 71.60; 78.70; S1; S4)
(a) Vol. **102**, No. 1, p. 47—65 (1987).

A. Seeger, J. Major, and F. Banhart
*The Study of Positron Diffusion in Solids by Positron Spin Relaxation (e^+SR) Experiments
(66.30 and 71.60; 75.50; S1.1)
(a) Vol. **102**, No. 1, p. 91—106 (1987).

A. Belaidi and R. N. West
*The Effects of Lattice Anisotropy on Positron Parameters in In–Pb Alloys
(61.55 and 71.60; 78.70; S4)
(a) Vol. **102**, No. 1, p. 107—112 (1987).

C. Hidalgo, S. Linderoth, and N. de Diego
*Positron–Dislocation Interaction in Deformed Zn and Cd
(71.60; 61.70; S1; S1.2)
(a) Vol. **102**, No. 1, p. 113—117 (1987).

A. Seeger and F. Banhart
*On the Systematics of Positron Lifetimes in Metals
(71.60; 61.70; S2; S5)
(a) Vol. **102**, No. 1, p. 171—179 (1987).

H.-E. Schaefer and F. Banhart
Thermal Equilibrium Vacancies in Platinum Studied by Positron Annihilation
(61.70; 71.60; 78.70; S1.3)
(a) Vol. **104**, No. 1, p. 263—272 (1987).

71.70. Level splitting and interactions

(see 75.10. — in magnetic phenomena, 75.30. Exchange and superexchange interactions, 73.20. Electronic surface states)

P. P. Bodiul, V. F. Garabazhiu, and D. V. Gitsu
The Impurity States in Bismuth
(71.55; 71.70; S4)
(b) Vol. **139**, No. 1, p. 245—256 (1987).

Du Mao-Lu and Zhao Min-Guang
*A Theoretical Explanation for Pressure Shift of Optical Spectra in Chromium Potassium Alum
(71.70 and 78.40; S11)
(b) Vol. **139**, No. 1, p. K87—K91 (1987).

D. Neogy and T. Purohit
*The Behavior of Active Centers in a Laser Host. A Crystal Field Investigation of Sm^{3+} in LaF_3 Single Crystal
(71.70; 65.40; 75.30; S9)
(b) Vol. **139**, No. 2, p. 519—525 (1987).

Yu. M. Seidov and A. M. Suleimanov
*To the Theory of Two-Magnon Absorption of Light by Surface Collective Excitations in the Singlet–Triplet Ferromagnets
(71.70; 75.30; 78.20; S1.4)
(b) Vol. **140**, No. 1, p. 149—154 (1987).

E. Schultheiss, A. Scharmann, and D. Schwabe
*Energy Levels of Triply Ionized Praseodymium in $BiLiF_4$ Crystals
(71.70 and 78.40; 63.20; S9.16)
(b) Vol. **140**, No. 1, p. 173—189 (1987).

S. M. Dubiel
*Changes in the ^{119}Sn-Site Hyperfine Field of Iron Induced by Substitutional Vanadium and Chromium
(71.70; 76.80; S1.1)
(b) Vol. **140**, No. 1, p. 191—202 (1987).

Yu Wan-Lun and Zhao Min-Guang
*Zero-Field Splitting and the d–d Transitions of Mn^{2+} on Ca(II) Sites in $Ca_5(PO_4)_3F$
(71.70; S11)
(b) Vol. **140**, No. 1, p. 203—212 (1987).

J.-C. Krupa, C. Khan Malek, P. Delamoye, B. Moine, and C. Pedrini
Observation of an Incommensurate Structure by Optical Spectroscopy with a 5f Ion as an Internal Probe. U^{4+}–$ThCl_4$ System
(78.50 and 78.55; 71.70; S9)
(b) Vol. **140**, No. 1, p. 289—300 (1987).

M. Wöhlecke, J. Schöpp, U. Scherz, and A. Jovanović
*Invariant Harmonic Oscillator Functions. III. The Cubic Groups T and T_h
(63.20 and 71.70)
(b) Vol. **141**, No. 2, p. 415—420 (1987).

J. SCHÖPP, M. WÖHLECKE, and U. SCHERZ
*Invariant Harmonic Oscillator Functions. IV. The Thirty Two Point Groups
(63.20 and 71.70)
(b) Vol. **143**, No. 1, p. 91—98 (1987).

WEN-CHEN ZHENG
*Theoretical Calculations of EPR Parameters and Absorption Spectra in Chromium-Doped α-$LiIO_3$ Crystals and Determination of Local Compressibilities
(71.70; 62.20; 76.30; 78.50; S11.1)
(b) Vol. **143**, No. 1, p. 217—223 (1987).

JIA-JUN CHEN and MIN-GUANG ZHAO
*High-Order Perturbation Formulae of the Spin-Hamiltonian Parameters for the Ground State $^3A_2(F)$ of d^8 Ions in Rhombic Symmetry
(71.70; S9.12)
(b) Vol. **143**, No. 2, p. 647—653 (1987).

SHANG BO, HUANG MING, and MIN-GUANG ZHAO
*The Relationship between the Pressure and Zero-Field Splitting of $ZnSiF_6 \cdot 6\,H_2O:Ni^{2+}$ and the Determination of Its Equations of State
(71.70; S9)
(b) Vol. **144**, No. 1, p. K35—K38 (1987).

T. S. N. MURTHY and M. SALAGRAM
*Electronic Spectra of Radiation-Induced $CrO^3{}_4{}^-$ in $3\,CdSO_4 \cdot 8\,H_2O$ Crystals
(61.80 and 71.70; 78.50; S11)
(b) Vol. **144**, No. 2, p. 519—527 (1987).

T. F. VEREMEICHIK
*The Regularities in Ground- and Excited-State Absorption Spectra of Impurity Ions of the Iron Group in Crystals
(71.70 and 78.50)
(b) Vol. **144**, No. 2, p. 709—717 (1987).

M. OKADA, T. KAWAKUBO, T. SEIYAMA, and M. NAKAGAWA
Enhancement of 3d-Electron Transitions in Neutron-Irradiated $MgO:Mn^{2+}$ Crystals
(78.50; 71.70; S10.1)
(b) Vol. **144**, No. 2, p. 903—909 (1987).

G. FISCHER and A. HERR
*Eigenvectors of $J = 5/2$ Ion States in a Tetragonal Crystal Field, when a Magnetic Field is Present in the [100] and [110] Directions as a Perturbation
(71.70)
(b) Vol. **144**, No. 2, p. K105—K108 (1987).

N. GUSKOS
An Interpretation of the Electric Field Effect on Cubic Gd^{3+} Centers in BaF_2 Single Crystals
(76.30; 71.70; S9.12)
(b) Vol. **144**, No. 2, p. K139—K142 (1987).

72. Electronic transport in condensed matter
(for surfaces, interfaces, and thin films, see 73)

72.10. Theory of electronic transport; scattering mechanism

F. Goedsche
*Modifications of the Hopping Conductivity Due to Long-Range Coulomb Interactions
(72.10 and 72.20; 71.30)
(b) Vol. **140**, No. 1, p. 225—231 (1987).

A. Dargys and A. F. Rudolph
*Tunneling in *k*-Space. Application to Degenerate Valence Band in the Infinite Spin–Orbit Interaction Limit
(72.10; 71.25)
(b) Vol. **140**, No. 2, p. 535—544 (1987).

Yu. B. Gaididei and I. I. Yakimenkó
*On Trigger Properties of Molecular Organizates. I. Tunnel Transport in Donor–Acceptor Systems with Strong Electron–Deformation Interaction
(72.10; 73.40)
(b) Vol. **140**, No. 2, p. 559—568 (1987).

Partha Pratim Basu, C. K. Sarkar, and D. Chattopadhyay
*Ionised-Impurity-Scattering Mobility in Semiconductors with Broadened Density of States
(72.10 and 72.20)
(b) Vol. **140**, No. 2, p. K129—K132 (1987).

Yu. B. Gaididei and I. I. Yakimenko
*On Trigger Properties of Molecular Organizates. II. Switch Waves in Donor–Acceptor Systems with Strong Electron–Deformation Interaction
(72.10; 73.40)
(b) Vol. **141**, No. 1, p. 213—220 (1987).

B. R. Bulka
*Quantum Interference in a Series of Metallic Rings in a Magnetic Field
(72.10)
(b) Vol. **141**, No. 1, p. 239—246 (1987).

C. K. Sarkar and P. K. Basu
*Electron Mobility in Quantum Well Structures Due to Alloy Disorder
(72.10)
(b) Vol. **141**, No. 2, p. K113—K115 (1987).

Ž. Kancleris and A. Matulis
*Two-Particle Monte-Carlo Method for Warm Electrons in a High-Frequency Electric Field
(72.10; 72.20; S5.12)
(b) Vol. **142**, No. 1, p. 247—254 (1987).

W. Hergert, S. Hildebrandt, and L. Pasemann
Theoretical Investigations of Combined EBIC, LBIC, CL, and PL Experiments. The Information Depth of the PL Signal
(78.55 and 78.60; 72.10)
(a) Vol. **102**, No. 2, p. 819—828 (1987).

T. T. MNATSAKANOV
*Transport Coefficients and Einstein Relation in a High Density Plasma of Solids
(72.10 and 72.30)
(b) Vol. **143**, No. 1, p. 225—234 (1987).

I. A. CHAIKOVSKII and A. I. GERMAN
*Sound Absorption in Highly Inhomogeneous Semiconductors
(72.10)
(b) Vol. **143**, No. 1, p. K53—K56 (1987).

A. DARGYS
*Tunnelling in *k*-Space. The Case of Time-Varying Electric Field
(72.10; 71.25)
(b) Vol. **143**, No. 2, p. 675—687 (1987).

72.15. Electronic conduction in metals and alloys

H.-J. FÖCKEL
Stationary and Resonance States at the Core of a Screw Dislocation and Their Contributions to Electrical Resistivity
(61.70 and 71.55; 72.15; S2)
(b) Vol. **139**, No. 1, p. 59—66 (1987).

M. INOUE, M. KOYANO, K. FUKUSHIMA, H. NEGISHI, and M. SASAKI
*Pressure Dependence of Electrical Resistivity in Intercalation Compounds M_xTiS_2 (M = 3d Transition Metal)
(72.15 and 72.20; S8)
(b) Vol. **139**, No. 1, p. 273—280 (1987).

T. IKARI, H. BERGER, and F. LEVY
*Electrical Properties of $NbTe_4$ and $TaTe_4$
(72.15; S1.63)
(b) Vol. **139**, No. 1, p. K37—K40 (1987).

A. ROJEK and C. SUŁKOWSKI
*Anomalous Magnetic and Electrical Properties of the $EuRh_{1.5}Sn_{4.2}$ Compound
(72.15 and 75.60; S1.4)
(b) Vol. **139**, No. 1, p. K63—K67 (1987).

J. M. BARANDIARÁN, J. C. GÓMEZ-SAL, J. RODRÍGUEZ FERNÁNDEZ, R. J. LÓPEZ SÁNCHEZ, and O. V. NIELSEN
*Temperature Dependence of the Electrical Resistivity in $[Co_{1-x}(Fe_{0.5}Ni_{0.5})_x]_{75}Si_{15}B_{10}$ Metallic Glasses
(72.15; S1.1; S1.2)
(a) Vol. **99**, No. 1, p. 243—249 (1987).

L. A. HUSSAIN and O. AKINLADE
*A Model Potential and Its Application
(71.45 and 72.15; 75.20)
(b) Vol. **139**, No. 2, p. K123—K128 (1987).

M. MEHAR KHAN and S. M. RAFIQUE
*Electronic Transport Properties of Gallium: Harrison's Method
(72.15; S4)
(b) Vol. **139**, No. 2, p. K131—K135 (1987).

M. M. AMADO, J. D. MONTENEGRO, J. M. MOREIRA, M. E. BRAGA, and J. B. SOUSA
Study of the Magnetic Transitions in $CeFe_2$ with High Resolution Transport Property Measurements
(75.40; 72.15; 75.50; S1.1; S1.4)
(a) Vol. **99**, No. 2, p. 625—630 (1987).

Y. IGASAKI and H. MITSUHASHI
*Polynomial Expression of the Bloch-Grüneisen Integral — Application to an Analysis of the Resistivity–Temperature Variation of Metals
(72.15)
(a) Vol. **99**, No. 2, p. K111—K115 (1987).
Erratum
(a) Vol. **101**, No. 2, p. K179 (1987).

T. SATOW, O. UEMURA, K. MATSUMOTO, and S. OKAMURA
*Electronic Properties of Liquid Sb–Se and Bi–Se Alloys
(72.15; 61.25; 75.20; S4)
(b) Vol. **140**, No. 1, p. 233—242 (1987).

A. KUMAR, G. CHANDRA, and O. P. KATYAL
*The Effect of Grain Boundary Scattering on the Electrical Resistivity of Polycrystalline Lead, Tin, and Tin–Lead Alloy Films
(72.15 and 73.60; 78.65; S4)
(a) Vol. **100**, No. 1, p. 169—175 (1987).

S. U. JEN and Y. D. YAO
*Electrical Resistivity of Ordered $Fe_{3-x}Pt_{1+x}$ Alloys between 4 and 1200 K
(72.15; 64.70; S1.1)
(a) Vol. **100**, No. 1, p. 269—275 (1987).

A. WOJAKOWSKI, P. J. MARKOWSKI, Z. HENKIE, and CH. LAURENT
*Anisotropy of the Electrical Resistivity in the Magnetically Uniaxial Compound UAsS
(72.15; S1.63)
(a) Vol. **100**, No. 1, p. K47—K49 (1987).

R. B. LAL, J. H. DAVIS, and H. U. WALTER
*Anisotropic Transport Properties of InBi
(72.15; S4)
(a) Vol. **100**, No. 2, p. 583—588 (1987).

P. Kopčanský
*The Electrical Resistivity of the Amorphous Fe–Ni–B System after Neutron Irradiation — Model of a Spinodal Decomposition
(72.15; S1.1)
(a) Vol. **100**, No. 2, p. K159—K163 (1987).

N. E. Alekseevskii, A. V. Mitin, E. P. Khlybov, A. Gilewski, and B. Greń
*The Influence of the "Quality" of the Specimens on the Galvanomagnetic Properties of UBe_{13}
(72.15; S1.5)
(a) Vol. **100**, No. 2, p. K165—K168 (1987).

B. Lippold, H. Müller, and S. Mahmoud
*Quadrupole Scattering in $PrNi_5$
(72.15; S1.2; S1.4)
(b) Vol. **141**, No. 1, p. 247—251 (1987).

J. Mucha, K. Gałuszewski, A. Jeżowski, and J. Rafałowicz
*Thermal Conductivity Minimum, Electrical Resistivity, and Lorenz Function of Zinc Monocrystals
(72.15; S1.2)
(b) Vol. **142**, No. 1, p. K57—K61 (1987).

A. Sakakibara, M. Yamada, T. Kanadani, and M. Ohta
*Effect of Pre-Aging on the Aging of Al–Zn Alloys
(72.15; S4)
(a) Vol. **102**, No. 1, p. K33—K35 (1987).

E. Kolley and W. Kolley
Mobility Edges for Interacting Electrons Based on Hubbard's Alloy Analogy
(71.30; 72.15)
(b) Vol. **142**, No. 2, p. 481—492 (1987).

C. Chaïb and J.-G. Gasser
*The Temperature Dependence of the Resistivity of Liquid Copper
(72.15; 61.25; S1.2)
(b) Vol. **142**, No. 2, p. 531—536 (1987).

M. Kurisu, H. Kadomatsu, and H. Fujiwara
Magnetic Field Dependences of Two Kinds of Structural Transition Temperatures of $CeAg_{1-x}In_x$
(64.60 and 75.30; 72.15; S1.3; S1.4)
(b) Vol. **142**, No. 2, p. 581—585 (1987).

P. V. Reddy and D. Akhtar
*Electrical Transport Properties of Amorphous $Ni_{60}Nb_{40-x}Cr_x$ Alloys
(72.15; S1.2)
(a) Vol. **102**, No. 2, p. K95—K98 (1987).

U. Krey, H. Ostermeier, and J. Zweck
Electronic Properties of Glassy Metals
(71.20; 72.15; S1.1; S1.2)
(b) Vol. **144**, No. 1, p. 203—212 (1987).

I. A. Schwirtlich, J. Petzold, and H. Schultz
*Resistivity Recovery in Niobium and Tantalum Following Electron Irradiation near 5 and 8 K
(61.80 and 72.15; S1)
(a) Vol. **104**, No. 1, p. 297—314 (1987).

A. Bartels, D. Bartusel, and K. Lücke
On the Nature of Frenkel Defects in Concentrated AuAg Alloys
(61.70; 72.15; S1.3)
(a) Vol. **104**, No. 1, p. 315—328 (1987).

P. Linden, P. Lamparter, and S. Steeb
Crystallization and Relaxation of Amorphous Mg–Zn, Cu–Ti, and Ni–P Alloys under High Hydrostatic Pressure
(61.40 and 62.50; 72.15; S1.2; S3)
(a) Vol. **104**, No. 1, p. 357—368 (1987).

G. J. Vázquez and L. F. Magaña
*Ab-initio Calculation of the Volume Dependence of the Resistivity of Aluminium
(72.15; 63.20; S4)
(b) Vol. **144**, No. 2, p. K119—K123 (1987).

T. Tsuchida and R. Kaneko
Temperature Dependence of Activation Energy in Stage II Recovery in Deformed Aluminium
(61.70; 72.15; S4)
(a) Vol. **104**, No. 2, p. 627—634 (1987).

H. Ratajczak, I. Gościańska, and A. Szlaferek
The Influence of Isothermal Annealing on Resistivity of Amorphous $Fe_{40}Ni_{40}B_{20}$ Thin Films
(61.40 and 68.55; 72.15; S1.1)
(a) Vol. **104**, No. 2, p. 731—741 (1987).

H. Matsui
*Characteristic Temperatures of Uranium Monophosphide (UP) in Magnetic Susceptibility and Electrical Resistivity
(72.15 and 75.3; S1.62)
(a) Vol. **104**, No. 2, p. 825—831 (1987).

72.20. Conductivity phenomena in semiconductors and insulators
(for nonelectric conduction, see 66.70)

P. J. S. Foot, T. Katz, S. N. Patel, B. A. Nevett, A. R. Piercy, and A. A. Balchin
*The Structures and Conduction Mechanisms of Lithium-Intercalated and Lithium-Substituted Nickel Phosphorus Trisulphide ($NiPS_3$), and the Use of the Material as a Secondary Battery Electrode **(Review Article)**
(72.20; 61.60; S8.16)
(a) Vol. **100**, No. 1, p. 11—29 (1987).

O. A. Golikova
*Semiconductors with Complex Lattice and the Amorphization Problem **(Review Article)**
(61.40 and 72.20; S5)
(a) Vol. **101**, No. 2, p. 277—314 (1987).

K. P. Ghatak and M. Mondal
Theoretical Analysis of the Oscillatory Diffusivity–Mobility Ratio in Degenerate n-InSb in the Presence of a Quantizing Magnetic Field
(71.25; 72.20; S7.13)
(b) Vol. **139**, No. 1, p. 195—203 (1987).

S. S. Kubakaddi, B. G. Mulimani, and V. M. Jali
*Quantum Theory of Thermopower in Quasi-Two-Dimensional Semiconductor Quantum Well Structures
(72.20; 73.40)
(b) Vol. **139**, No. 1, p. 267—271 (1987).

M. Inoue, M. Koyano, K. Fukushima, H. Negishi, and M. Sasaki
*Pressure Dependence of Electrical Resistivity in Intercalation Compounds M_xTiS_2 (M = 3d Transition Metal)
(72.15 and 72.20; S8)
(b) Vol. **139**, No. 1, p. 273—280 (1987).

P. P. Petrov, V. Vassilev, and Z. Boncheva-Mladenova
*Transport Properties of Monocrystalline Silver Telluride
(72.20; S8)
(a) Vol. **99**, No. 1, p. 237—242 (1987).

W. Seifert and M. Kittler
Negative (Bright) EBIC Contrast at Oxygen-Induced Defects in Silicon
(61.70; 72.20; S5.11)
(a) Vol. **99**, No. 1, p. K11—K14 (1987).

F. Comas, C. Trallero Giner, and J. Tutor
*Ohmic Mobility of Quasi-One-Dimensional Semiconducting Structures. Acoustic and Piezoelectric Phonon Scattering
(72.20 and 73.60; S7.12; S8.11)
(b) Vol. **139**, No. 2, p. 433—440 (1987).

V. A. KASIYAN, D. D. NEDEOGLO, A. V. SIMASHKEVICH, and I. N. TIMCHENKO
*Magnetoresistivity of ZnSe in the Region of Hopping Conduction over Localized States in the Coulomb Gap
(72.20; 71.55; S8.12)
(b) Vol. **139**, No. 2, p. 559—564 (1987).

G. CH. CHRISTAKUDIS, S. K. PLACHKOVA, E. S. AVILOV, L. E. SHELIMOVA, and N. KH. ABRIKOSOV
*Electrical Resistivity and Scattering Mechanisms of $(GeTe)_{1-x}((Ag_2Te)_{1-y}(Sb_2Te_3)_y)_x$ Solid Solutions ($y = 0.6$)
(72.20; S8.15)
(a) Vol. **99**, No. 2, p. 593—601 (1987).

Z. DZIUBA
*Analysis of the Electrical Conduction in CdHgTe Crystals
(72.20; S8.15)
(b) Vol. **140**, No. 1, p. 213—223 (1987).

F. GOEDSCHE
*Modifications of the Hopping Conductivity Due to Long-Range Coulomb Interactions
(72.10 and 72.20; 71.30)
(b) Vol. **140**, No. 1, p. 225—231 (1987).

T. DRÜSEDAU, D. WEGENER, and R. BINDEMANN
*The Ordinary and Generalized Meyer-Neldel Rule in a-Si:H Calculated by the Statistical Fermi Level Shift
(71.20 and 72.20; S5.11)
(b) Vol. **140**, No. 1, p. K27—K31 (1987).

O. ZIEP and R. KEIPER
*Self Localization of Mobile Electrons in the Theory of Quantum Hall Effect
(72.20; 73.40)
(b) Vol. **140**, No. 1, p. K41—K46 (1987).

M. L. KOZHUKH and N. S. LIPKINA
The Electrical Conductivity of Disordered Germanium near the Metal–Insulator Transition
(72.60; 72.20; S5.12)
(a) Vol. **100**, No. 1, p. 259—267 (1987).

V. YA. SUKHAREV and I. A. MYASNIKOV
*On the Conductivity of a Disordered Semiconductor System with Grain Barriers under Active Particle Adsorption
(72.20; 73.60)
(a) Vol. **100**, No. 1, p. 277—282 (1987).

G. KÜHNEL, W. SIEGEL, and H. A. SCHNEIDER
*High-Field Current–Voltage Characteristics of Semi-Insulating GaAs
(72.20; S7.12)
(a) Vol. **100**, No. 1, p. 283—287 (1987).

G. KÜHNEL, W. SIEGEL, S. FELIX, and H. A. SCHNEIDER
*On the Determination of the Conductivity and the Hall Coefficient of Semi-Insulating GaAs at Room Temperature
(72.20; S7.12)
(a) Vol. **101**, No. 2, p. 531—536 (1987).

Ž. KANCLERIS and A. MATULIS
Two-Particle Monte-Carlo Method for Warm Electrons in a High-Frequency Electric Field
(72.10; 72.20; S5.12)
(b) Vol. **142**, No. 1, p. 247—254 (1987).

H.-U. MÜLLER, F. LUDWIG, and R. HERRMANN
*Cyclotron Resonance in the n-Inversion Layer of InSb Grain Boundaries
(72.20 and 76.40; S7.13)
(b) Vol. **142**, No. 1, p. K69—K71 (1987).

YU. P. DROZHOV
*Galvanomagnetic Effects in an Inhomogeneous Relaxation Semiconductor
(72.20)
(a) Vol. **102**, No. 1, p. 349—355 (1987).

G. CH. CHRISTAKUDIS, S. K. PLACHKOVA, and L. E. SHELIMOVA
*Thermoelectric Power of $(GeTe)_{1-x}((Ag_2Te)_{1-y}(Sb_2Te_3)_y)_x$ Solid Solutions ($y = 0.6$; $0 \leqq x \leqq 1$)
(72.20; S8.16)
(a) Vol. **102**, No. 1, p. 357—366 (1987).

C. DONOLATO and M. BIANCONI
Use of the EBIC Contrast Profile Area for Evaluating the Recombination Strength of Dislocations
(61.70; 72.20; S5.11)
(a) Vol. **102**, No. 1, p. K7—K11 (1987).

J. MOLENDA, A. STOKŁOSA, and W. ZNAMIROWSKI
*Transport Properties of Ferrous Oxide $Fe_{1-y}O$ at High Temperature
(72.20; 71.30; S10.1)
(b) Vol. **142**, No. 2, p. 517—529 (1987).

YU. P. DROZHOV
*Photo- and Field-Induced Random Field in Relaxation Case Semiconductors
(72.20 and 72.40)
(a) Vol. **102**, No. 2, p. 733—739 (1987).

K. KIBICKAS, J. PARŠELIŪNAS, and Š. VASILIAUSKAS
*Transient Hot Electron Phenomena in GaAs n^+–n–n^+ Structures at 300 K
(72.20 and 73.40; S7.12)
(a) Vol. **102**, No. 2, p. K99—K102 (1987).

V. M. JALI, S. S. KUBAKADDI, and B. G. MULIMANI
*On the Interpretation of Thermopower Measurements in GaAs–$Al_xGa_{1-x}As$
(72.20 and 73.40; S7.15)
(b) Vol. **143**, No. 1, p. K25—K29 (1987).

A. I. KASIYAN, A. G. CHEBAN, M. A. CHEBAN, I. V. SUR, and O. V. ZAKHARCHUK
*Effect of Traps on the Shape of the Output Signal of Ionizing-Radiation Semiconductor Detectors
(72.20 and 85; S5.11)
(a) Vol. **103**, No. 1, p. 185—191 (1987).

J. PANCÍŘ, J. HORÁK, and Z. STARÝ
*Concentration of Free Carriers in $Bi_{2-x}In_xTe_3$ Mixed Crystals
(72.20; S8.15)
(a) Vol. **103**, No. 2, p. 517—526 (1987).

V. M. JALI, S. S. KUBAKADDI, and B. G. MULIMANI
*Quantum Theory of Thermopower in Quantum Well Wires
(72.20; 73.40)
(b) Vol. **144**, No. 2, p. 739—744 (1987).

S. ABDALLA, M. DONGOL, and M. IBRAHIM
*Electrical Conduction in Amorphous $Cu_{10}As_{40}Se_{50}$ with Potential Fluctuations
(72.20; S8)
(b) Vol. **144**, No. 2, p. 745—752 (1987).

A. MÖBIUS
The Metal–Semiconductor Transition in Amorphous $Si_{1-x}Cr_x$ Films. Detection of the Transition by Room-Temperature Measurements
(71.30 and 72.60; 72.20; 73.60; S1.61)
(b) Vol. **144**, No. 2, p. 759—766 (1987).

P. P. BODIUL, M. P. BOIKO, D. V. GITSU, and N. A. REDKO
*Commutation Effect in the Magneto-Thermopower of $Bi_{0.88}Sb_{0.12}$ Alloy
(72.20; S4)
(b) Vol. **144**, No. 2, p. K109—K112 (1987).

D. A. ARONOV and B. R. MAMATKULOV
*The Influence of Field-Induced Variation in Majority Carrier Lifetime Caused by Photostimulated Heating on Impurity Photoconductivity of Compensated Semiconductors with Exclusion
(72.20 and 72.40; 73.40)
(a) Vol. **104**, No. 2, p. 785—792 (1987).

A. I. EATAH, A. A. GHANI, M. F. EL-SHAHAT, and E. EL-FARAMAWY
*Temperature Dependence of Thermoelectric Power and Electrical Resistivity in $Cu_xCo_{1-x}Fe_2O_4$ Ferrites
(72.20; S11.2)
(a) Vol. **104**, No. 2, p. 793—797 (1987).

72.30. High-frequency effects; plasma effects

B. N. NARAHARI ACHAR
*Non-Local Effects in Helicon Wave Propagation in a Periodic Structure
(72.30)
(b) Vol. **140**, No. 1, p. K37—K40 (1987).

M. ASCHE and O. G. SARBEI
Nonequilibrium Dynamics of Carriers and Phonons in GaAs after High Excitation by Short Time Laser Pulses
(71.38; 72.30; S7.12)
(b) Vol. **141**, No. 2, p. 487—491 (1987).

T. T. MNATSAKANOV
*Transport Coefficients and Einstein Relation in a High Density Plasma of Solids
(72.10 and 72.30)
(b) Vol. **143**, No. 1, p. 225—234 (1987).

B. N. NARAHARI ACHAR
*Helicon Waves in a Sinusoidal Periodic Structure
(72.30)
(b) Vol. **143**, No. 1, p. 235—242 (1987).

S. GHOSH, R. B. SAXENA, and M. S. PATHAK
*Modulation of a High-Power Helicon Wave in Longitudinally Magnetised Semiconducting Plasmas
(72.30)
(a) Vol. **104**, No. 1, p. K53—K56 (1987).

72.40. Photoconduction and photovoltaic effects; photodielectric effects

J. BULLOT and M. P. SCHMIDT
Physics of Amorphous Silicon–Carbon Alloys **(Review Article)**
(61.40; 71.20; 71.55; 72.40; 73.60; 78.20; 78.55; 78.60; S6)
(b) Vol. **143**, No. 2, p. 345—418 (1987).

N. V. UNNIKRISHNAN, H. S. BHATTI, and R. D. SINGH
*Two-Photon Photoconductivity in CdI_2 under Double Beam Laser Excitation
(72.40; S9)
(b) Vol. **139**, No. 1, p. K41—K45 (1987).

YU. P. DROZHOV
*Photoconductivity in Relaxation Case Semiconductors
(72.40)
(a) Vol. **100**, No. 2, p. K155—K158 (1987).

R. P. SINGH and S. L. SINGH
Photoelectrochemical Solar Cells Using Vacuum-Evaporated Zinc Phosphide Films
(73.40; 72.40; S7)
(a) Vol. **101**, No. 1, p. 129—135 (1987).

YU. P. DROZHOV
*Photoconductivity and Photovoltage in a Semiconductor in the Relaxation Case
(72.40)
(a) Vol. **101**, No. 1, p. 205—216 (1987).

C. POPESCU, T. STOICA, and A. G. SAVANÉ
An Equation for the Current–Voltage Curve of the a-Si:H Solar Cells
(73.40; 72.40; S5.11)
(a) Vol. **101**, No. 2, p. 469—477 (1987).

YU. P. DROZHOV
*Photogalvanomagnetic Effect in Relaxation Case Semiconductor
(72.20 and 72.40)
(a) Vol. **101**, No. 2, p. 515—520 (1987).

H. BERGNER, V. BRUECKNER, V. S. DNEPROVSKII, M. GRAUROCK, and M. LENZNER
*Investigation of Recombination Mechanisms in GaSe by Optoelectronic Switching
(72.40; S8)
(a) Vol. **101**, No. 2, p. K157—K160 (1987).

J. P. GUESDON, B. KOBBI, C. JULIEN, and M. BALKANSKI
*Electrical and Photovoltaic Properties of In_xSe_{1-x} Thin Films
(72.40 and 73.60; S8)
(a) Vol. **102**, No. 1, p. 327—335 (1987).

YU. P. DROZHOV
*Photo- and Field-Induced Random Field in Relaxation Case Semiconductors
(72.20 and 72.40)
(a) Vol. **102**, No. 2, p. 733—739 (1987).

D. I. TSIULYANU, E. P. KOLOMEYKO, and V. N. STAMOV
*Photoelctric Phenomena in Isotope Vitreous As_2S_3–Silicon Single-Crystal Heterojunctions
(72.40 and 73.40; S5.11; S8)
(a) Vol. **102**, No. 2, p. K103—K107 (1987).

T. WILSON and E. M. MCCABE
Photoluminescence and Optical Beam Induced Current Images of Defects in Semiconductors
(61.70; 72.40; 78.55)
(a) Vol. **103**, No. 1, p. 107—113 (1987).

H. Bergner, V. Brückner, F. Kerstan, M. Lenzner, and W. Nowick
*An Optoelectronic Switch with Increased Sensitivity Using a Finger Structure Electrode
(72.40; S5.11)
(a) Vol. **103**, No. 1, p. 193—197 (1987).

J. Pastrňák, F. Karel, and W. Ulrici
*Two-Light-Source Extrinsic Photoconductivity of GaAs:V
(72.40; 71.55; S7.12)
(a) Vol. **103**, No. 1, p. 205—212 (1987).

D. P. Joshi
*Grain Boundary Recombination in Polycrystalline Silicon under Optical Illumination
(72.40; 73.40; S5.11)
(a) Vol. **103**, No. 1, p. 213—218 (1987).

B. Kumar
*Carrier Drift Mobilities in Some Organic Solids. A Search for Organic Photovoltaic Materials. II. Charge Transport Calculations in s-Tetrazine and 4 Phenylazoazobenzene
(72.40 and 72.80; 71.25; S12)
(b) Vol. **143**, No. 2, p. 663—673 (1987).

K. Tennakone, A. H. Jayatissa, C. A. N. Fernando, S. Wickramanayake, S. Punchihewa, L. K. Weerasena, and W. D. R. Premasiri
Semiconducting and Photoelectrochemical Properties of n- and p-Type β-CuCNS
(73.60 and 78.65; 68.55; 72.40; S11)
(a) Vol. **103**, No. 2, p. 491—497 (1987).

K. Nakashima, S. Nojima, Y. Kawamura, and H. Asahi
Deep Electron Trapping Centers in Si-Doped InAlAs Grown by Molecular Beam Epitaxy
(71.55; 72.40; S7.15)
(a) Vol. **103**, No. 2, p. 511—516 (1987).

D. A. Aronov and B. R. Mamatkulov
*The Influence of Field-Induced Variation in Majority Carrier Lifetime Caused by Photostimulated Heating on Impurity Photoconductivity of Compensated Semiconductors with Exclusion
(72.20 and 72.40; 73.40)
(a) Vol. **104**, No. 2, p. 785—792 (1987).

P. Hertel, K. H. Ringhofer, and R. Sommerfeldt
*Theory of Thermal Hologram Fixing and Application to $LiNbO_3$:Cu
(72.40 and 78.20; S11.1)
(a) Vol. **104**, No. 2, p. 855—862 (1987).

E. Z. Katsnelson
*Photogalvanomagnetic Properties of Manganese Ferrite
(72.40; S11.2)
(a) Vol. **104**, No. 2, p. K127—K132 (1987).

G. J. Papaioannou, C. D. Kourkoutas, N. G. Theofanous, and A. T. Arapoyianni
*A Study of the Gain Mechanisms in GaAs Planar Optical Photoconductivive Detectors
(72.40 and 85; S7.12)
(a) Vol. **104**, No. 2, p. K141—K146 (1987).

72.50. Acoustoelectric effects

J. Ďurček, R. Fettig, and J. Rosenzweig
*The Charge Injection Effect on the Acousto-Electric Properties of Si–SiO_2 Structures
(72.50 and 73.40; S5.11; S10.1)
(a) Vol. **101**, No. 1, p. K33—K34 (1987).

A. V. Dobrynin, G. A. Naida, and V. A. Novoselov
*Acoustoelectric Parameters of Aluminium Nitride Epitaxial Layers
(72.50; S7.1)
(a) Vol. **104**, No. 1, p. K47—K51 (1987).

72.60. Mixed conductivity and conductivity transitions

M. L. Kozhukh and N. S. Lipkina
*The Electrical Conductivity of Disordered Germanium near the Metal–Insulator Transition
(72.60; 72.20; S5.12)
(a) Vol. **100**, No. 1, p. 259—267 (1987).

A. Möbius
*The Metal–Semiconductor Transition in Amorphous $Si_{1-x}Cr_x$ Films. Detection of the Transition by Room-Temperature Measurements
(71.30 and 72.60; 72.20; 73.60; S1.61)
(b) Vol. **144**, No. 2, p. 759—766 (1987).

72.70. Noise processes and phenomena

E. Kolley and W. Kolley
Stochastic Quantization Approach to Disordered Fermions
(71.10; 72.70)
(b) Vol. **140**, No. 2, p. 527—533 (1987).

72.80. Conductivity of specific semiconductors and insulators

B. Biglari, M. Samimi, J. M. Koebel, M. Hage-Ali, and P. Siffert
*Evolution of Resistivity Along Cl-Compensated THM Grown Cadmium Telluride Crystals
(72.80; S8.13)
(a) Vol. **100**, No. 2, p. 589—596 (1987).

D. Meimaris, J. Katris, D. Martakos, and G. P. Triberis
*The Effect of Thermal Annealing on the Electrical Conductivity of Bulk Cd–Ge–As Glass
(72.80; S7.16)
(a) Vol. **100**, No. 2, p. K169—K172 (1987).

W. Ulrici, L. Eaves, K. Friedland, and D. P. Halliday
Optical Studies of Vanadium in Gallium Phosphide
(71.55; 72.80; 78.50; 78.55; S7.11)
(b) Vol. **141**, No. 1, p. 191—202 (1987).

M. Wienecke and M. Schenk
In-Situ High-Temperature Hall and Conductivity Measurements for Investigations of the Stability Region of PbTe
(61.70; 66.30; 72.80; S8)
(a) Vol. **101**, No. 1, p. 115—122 (1987).

F. M. Kamarly and A. Yu. Yangirov
Magnetic Susceptibility of the $(SnTe)_{1-x}$–$(EuTe)_x$ System
(75.20; 72.80; S8.16)
(a) Vol. **101**, No. 1, p. K61—K63 (1987).

S. N. Toubektsis, J. A. Kalomiros, L. V. Giobliakis, K. M. Paraskevopoulos, A. N. Anagnostopoulos, and E. K. Polychroniadis
On the Growth and Characterization of the System $TlBiSe_2$–$TlBiTe_2$
(61.60; 72.80; 78.20; S8.16)
(a) Vol. **101**, No. 2, p. 355—360 (1987).

M. A. Hassan, C. A. Hogarth, and G. R. Moridi
*AC Conductivity of Copper Phosphate Glasses
(72.80; S11.1)
(a) Vol. **101**, No. 2, p. 537—542 (1987).

W. Siegel, H. Koi, G. Kühnel, and H. A. Schneider
Observation of the 78 meV Acceptor in p-Type GaAs Grown by the SSD Method
(71.55; 72.80; S7.12)
(a) Vol. **101**, No. 2, p. K147—K151 (1987).

A. Wojakowska and E. Kundys
*High Electrical Conductivity of a Solid Phase in the SnI_2–AgI System
(72.80; 64.70; S9.16)
(a) Vol. **101**, No. 2, p. K153—K155 (1987).

H. Kitagawa and H. Nakashima
Nickel-Related Donor Level in Silicon
(71.55; 72.80; S5.11)
(a) Vol. **102**, No. 1, p. K23—K27 (1987).

72.90. Other topics in electronic transport in condensed matter

F. Berz and J. A. Morice
*Accuracy of the Transmission Coefficient across Parabolic Barriers as Obtained from a Generalized WKB Approach
(72.90; 73.90)
(b) Vol. **139**, No. 2, p. 573—582 (1987).

P. Enders
Photon Recycling in Double Heterostructures. III. Variational Calculations
(78.55; 72.90; 73.40)
(b) Vol. **141**, No. 1, p. 317—323 (1987).

O. Ziep and R. Keiper
*Liquid Site Model and Self-Consistent Potentials in the Theory of Quantum Hall Effect
(72.90)
(b) Vol. **144**, No. 2, p. 767—779 (1987).

73. Electronic structure and electrical properties of surfaces, interfaces, and thin films

73.20. Electronic surface states
(for emission and impact phenomena, see 79)

A. L. Ivanovskii, D. L. Novikov, and V. A. Gubanov
Electronic Properties of Carbides, Nitrides, and Oxides of Subgroup VIa Transition Metals. Theory and Experiment **(Review Article)**
(71.25; 73.20; 78.70; 79.60; S1.61; S1.62; S10)
(b) Vol. **141**, No. 1, p. 9—33 (1987).

A. A. Karpushin, A. N. Sorokin, I. Yu. Semenova, and M. Tomášek
*A Contribution to the Theoretical Interpretation of the UPS Spectra of Monoatomic Adsorption of Oxygen on the Si (111) Surface
(68.20 and 73.20; 79.60; S5.11)
(b) Vol. **139**, No. 1, p. 125—130 (1987).

Tran Thoai Duy Bao
*Plasmon Excitation in the Interaction of Fast Electrons with Metal Surfaces
(71.45 and 73.20; 61.80; S4)
(b) Vol. **139**, No. 1, p. 145—154 (1987).

M. S. El-Dessouki, V. A. Attia, M. M. Saad El-Deen, and F. Z. Gobrial
*Surface Photovoltage Spectroscopy of Real ⟨111⟩ GaP Surfaces
(73.20 and 73.25; S7.11)
(a) Vol. **99**, No. 1, p. 171—179 (1987).

V. Filipavičius, R. Gaidys, V.-A. Matulaitis, G. Petrauskas, A. Sakalas, and S. Sakalauskas
*Investigation of the Surface States in Heavily Doped GaAs by Kelvin Probe
(73.20; S7.12)
(a) Vol. **99**, No. 2, p. 543—547 (1987).

Tran Thoai Duy Bao
*Image Force and Stopping Power of a Fast Moving Ion near a Metal Surface
(61.80 and 73.20; 71.45; S4)
(b) Vol. **140**, No. 2, p. 449—457 (1987).

M. Stęślicka and L. Jurczyszyn
*Virtual Image- and Field-Induced Electronic States
(73.20)
(b) Vol. **140**, No. 2, p. K99—K102 (1987).

K. P. Ghatak and S. Biswas
*Theoretical Analysis of the Thermoelectric Power in p-Channel Inversion Layers on Si under Strong Magnetic Quantization
(72.20 and 73.20; S5.11)
(b) Vol. **140**, No. 2, p. K107—K111 (1987).

M. Bhattacharya, S. Mukhopadhyay, and T. K. Mitra
*Surface Optical Polaron in Magnetic Field
(71.38 and 73.20; 68.30)
(b) Vol. **142**, No. 1, p. 141—148 (1987).

W. Lorenz and T. Katterle
*Chemical Processes on Semiconductor Electrode Interfaces and Distribution of Charge Transfer on the Orbital Energy Scale
(71.20 and 73.20; S7.12)
(b) Vol. **142**, No. 1, p. 149—163 (1987).

A. V. Vakulenko, R. A. Vardanyan, and L. B. Hovakimyan
*On the Luminescence at the Surface and in the Surface Space Charge Region
(73.20 and 78.55)
(b) Vol. **142**, No. 2, p. 417—422 (1987).

A. Beckmann
*On the Theory of Photoemission from Adsorbate-Induced Surface States
(73.20 and 79.60; S1.2)
(b) Vol. **144**, No. 1, p. 213—224 (1987).

J. Braun, G. Thörner, and G. Borstel
Relativistic Theory of Photoemission from Solids. III. Bulk and Surface Transition Matrix Elements
(79.60; 73.20)
(b) Vol. **144**, No. 2, p. 609—616 (1987).

P. Rennert and F. Seifert
Scattering Properties of Non-Spherical Muffin-Tin Potentials of Surface Atoms
(68.20 and 71.10; 73.20; S1.2)
(b) Vol. **144**, No. 2, p. 617—629 (1987).

G. Ghibaudo
*Calculation of Surface Charge Noise at the Si–SiO_2 Interface
(73.20 and 73.40; S5.11)
(a) Vol. **104**, No. 2, p. 917—930 (1987).

E. Chryssou and P. Vartzeli-Nikaki
*Exoelectron Emission (EE) of Heat Induced $SrSO_4$/SrS Conversion
(73.20 and 79.75; S11.1)
(a) Vol. **104**, No. 2, p. K111—K116 (1987).

73.25. Surface conductivity

V. I. Ivanov-Omskii, N. N. Berchenko, and A. I. Elizarov
Transport Anomalies in $Cd_xHg_{1-x}Te$ **(Review Article)**
(72.80; 73.25; S8.15)
(a) Vol. **103**, No. 1, p. 11—28 (1987).

M. S. El-Dessouki, V. A. Attia, M. M. Saad El-Deen, and F. Z. Gobrial
*Surface Photovoltage Spectroscopy of Real ⟨111⟩ GaP Surfaces
(73.20 and 73.25; S7.11)
(a) Vol. **99**, No. 1, p. 171—179 (1987).

J. A. Giacometti and R. Coelho
*Further Speculations on the Natural Decay of Corona Charged Insulating Films
(73.25; S12)
(a) Vol. **99**, No. 1, p. K45—K48 (1987).

Ch. Schnittler and S. Scherf
Alternative Quasi-Classical Transport Equations. II. Boltzmann's Equation and the Electron Distribution Function for MIS Systems
(73.40; 73.25; S5.11)
(b) Vol. **143**, No. 1, p. 325—333 (1987).

E. Gerlach
*An Estimate of Transport in Surface State Bands
(73.25)
(b) Vol. **143**, No. 1, p. K19—K24 (1987).

S. Scherf, Ch. Schnittler, and G. Paasch
Alternative Quasi-Classical Transport Equations. III. Charge Transport in MIS Systems with Inclusion of Diffuse Surface Scattering and Quantum Size Effects
(73.40; 73.25; S5.11)
(b) Vol. **143**, No. 2, p. 795—805 (1987).

73.30. Surface double layers, Schottky barriers, and work functions

K. Schmalz, K. Tittelbach-Helmrich, and H. Richter
*On DLTS-Measurements of the Barrier-Limited Capture Process
(71.55 and 73.30; S5.11)
(a) Vol. **99**, No. 2, p. K97—K102 (1987).

S. Saito and T. Maeda
*Temperature Variation of the Work Function of Polycrystalline Cobalt in the Temperature Range 100 °C $\leqq T \leqq$ 485 °C
(73.30; S1.2)
(a) Vol. **100**, No. 1, p. K21—K24 (1987).

H. LEMKE
*Symmetrieeigenschaften des Potentialverlaufes bei Schottky-Randschichten
(Symmetry Properties of the Potential Distribution in Schottky Junctions)
(73.30)
(a) Vol. **100**, No. 1, p. K31—K34 (1987).

I. E. TÜRE, A. W. BRINKMAN, G. J. RUSSELL, and J. WOODS
The Effects of Oxygen in Au–CdSe Schottky Diodes
(85; 68.20; 73.30; S8.12)
(a) Vol. **100**, No. 2, p. 681—691 (1987).

H. P. HINOV
*DC Voltage-Controlled Pulsations of a Homotropic Nematic Layer with a Free-Sliping Liquid Crystal–Polar Liquid Boundary
(61.30 and 73.30)
(a) Vol. **101**, No. 1, p. 185—192 (1987).

N. BALASUBRAMANYAM and VIKRAM KUMAR
*Pressure Dependence of Barrier Heights of Schottky Contacts on Silicon
(73.30; S5.11)
(a) Vol. **101**, No. 1, p. K29—K32 (1987).

H. LEMKE
*Eine analytische Approximation für den Potentialverlauf bei Schottky-Randschichten
(An Analytic Approximation for the Potential Distribution of Schottky Junctions)
(73.30)
(a) Vol. **101**, No. 1, p. K39—K43 (1987).

S. M. PATEL and V. G. KAPALE
*$AgGaSe_2$/Al Thin Film Schottky Barrier Diodes
(73.30 and 85; S8.16)
(a) Vol. **101**, No. 1, K87—K92 (1987).

V. B. BIKBAEV, S. Č. KARPINSKAS, V. E. STOŠKUS, and J. J. VAITKUS
*Forward Thermionic Emission Current of an Inhomogeneous Contact to Cadmium Chalcogenide
(73.30; S8.1)
(a) Vol. **102**, No. 2, p. 719—724 (1987).

H. LEMKE
*Eine exakte Lösung für den Potentialverlauf bei Schottky-Randschichten
(An Exact Solution for the Potential Distribution at Schottky Junctions)
(73.30)
(a) Vol. **103**, No. 1, p. K27—K32 (1987).

W. LORENZ
Chemical Bond Formation Processes on Semiconductor Interfaces at Electronic Equilibrium or Non-equilibrium
(68.45 and 73.40; 71.20; 73.30)
(b) Vol. **144**, No. 2, p. 585—599 (1987).

A. B. MAITY, C. MAJUMDAR, and A. N. CHAKRAVARTI
*On Photoemission from Quasi-One-Dimensional Structures of Narrow-Gap Semiconductors Having Rectangular Cross-Sections
(73.30 and 79.60; S7.13)
(b) Vol. **144**, No. 2, p. K93—K98 (1987).

73.40. Interfaces

S. S. KUBAKADDI, B. G. MULIMANI, and V. M. JALI
Quantum Theory of Thermopower in Quasi-Two-Dimensional Semiconductor Quantum Well Structures
(72.20; 73.40)
(b) Vol. **139**, No. 1, p. 267—271 (1987).

P. ENDERS
*Minority-Carrier Leakage over Step p–n Heterojunctions
(73.40; S7.12)
(a) Vol. **99**, No. 1, p. 181—192 (1987).

J.-P. ZÖLLNER, G. PAASCH, and H. ÜBENSEE
*The Influence of Quantum Effects on the Semiconductor lf- and hf-Capacitance
(73.40; S5.11)
(a) Vol. **99**, No. 1, p. 193—204 (1987).

A. SRIVASTAVA
*Threshold Voltage Dependence on Channel Length in Small Geometry MOSFETs
(73.40 and 85)
(a) Vol. **99**, No. 1, p. 303—308 (1987).

H. K. PHAN, L. H. PHU, and N. T. NGUYET ANH
*Switching Characteristics of MISS Devices Containing Electron Surface Traps
(73.40 and 85; S5.11)
(a) Vol. **99**, No. 1, p. 309—314 (1987).

B. RHEINLÄNDER, J. KOVÁC, P. NEVERMANN, and H. FIESELER
*On the Determination of Quantum Yield in InGaAsP Double-Heterostructure LED
(73.40 and 85; S7.15)
(a) Vol. **99**, No. 1, p. 315—320 (1987).

P. ENDERS
*Schrödinger Equation and Wave-Function Matching Conditions for Spatially Varying Effective Mass
(71.25 and 73.40)
(b) Vol. **139**, No. 2, p. K113—K116 (1987).

E. G. SALMAN, G. V. PEROV, and V. N. VERTOPRAKHOV
*Ionic Current Study in Thermally Grown Silicon Dioxide Films on Polycrystalline Silicon
(73.40 and 77.30; 77.55; S10.1)
(a) Vol. **99**, No. 2, p. 537—542 (1987).

M. KITTLER and W. SEIFERT
*Characterization of the Leakage-Current Diffusion Component in Intrinsically Gettered Silicon by EBIC
(73.40; 72.40; 85; S5.11)
(a) Vol. **99**, No. 2, p. 559—572 (1987).

M. PROCOP, H. RAIDT, and B. SANDOW
Auger Electron and X-Ray Spectroscopic Analysis of Au–Ge Contacts to GaAs
(68.48; 73.40; 79.20; S1.3; S7.12)
(a) Vol. **99**, No. 2, p. 573—580 (1987).

F. JANUSCHEWSKI and H.-J. ERZGRÄBER
*Modelling the Influence of Hot Electrons on the Transfer Characteristic of Short-Channel MOSFETs
(73.40 and 85)
(a) Vol. **99**, No. 2, p. 649—656 (1987).

H. HAEFNER, M. HEIDER, F. KOCH, R. MITDANK, G. OELGART, W. THIELEMANN, B. RHEINLÄNDER, and B. JACOBS
*Properties of Red Light-Emitting (AlGa)As Single-Heterostructure Diodes. I. Structure Characterization
(73.40 and 85; 78.55; 78.60; S7.15)
(a) Vol. **99**, No. 2, p. 657—667 (1987).

I. THURZO and E. PINČÍK
*Attempts to Find a Correlation between the GaAs Plasma Oxide Growth Conditions and the Charge DLTS Response of Interface States
(73.40; 71.55; S7.12)
(a) Vol. **99**, No. 2, p. K91—K95 (1987).

J. SEKULSKI
*On the Possibility of Application of the Derivative Lightoutput–Voltage Characteristic for the Analysis of Electron–Phonon Interaction in Electroluminescent ZnSe MOS Structures
(73.40 and 78.60; 8.12)
(a) Vol. **99**, No. 2, p. K139—K141 (1987).

N. T. TRAN and F. R. JEFFREY
*Effects of Carbon Profiles at the p/i Interface in Amorphous Silicon Solar Cells
(73.40 and 85; S5.11; S6)
(a) Vol. **99**, No. 2, p. K143—K148 (1987).

G. GHIBAUDO
*A Simple Model of the Drain Saturation Voltage Dependence with Gate Voltage for Short-Channel MOSFETs
(73.40 and 85)
(a) Vol. **99**, No. 2, p. K149—K153 (1987).

J. LÓPEZ-GONDAR, H. RODRIGUEZ-COPPOLA, and R. PÉREZ-ALVAREZ
*GaAs–AlGaAs Quantum Well of Compositional Graded Interface in the Envelope Function Approximation
(73.40; S7.12; S7.15)
(b) Vol. **140**, No. 1, p. 97—102 (1987).

A. J. ABU EL-HAIJA and A. H. MADJID
*DC Electrical Conductivity of Ultrathin Ag–SiO Superlattices
(73.40; S1.3; S10.1)
(b) Vol. **140**, No. 1, p. 113—120 (1987).

O. ZIEP and R. KEIPER
Self Localization of Mobile Electrons in the Theory of Quantum Hall Effect
(72.20; 73.40)
(b) Vol. **140**, No. 1, p. K41—K46 (1987).

I. Zólomy
*A Comprehensive Theory of MISS Devices Including Punch-Through and Avalanche-Induced Switchings
(73.40 and 85)
(a) Vol. **100**, No. 2, p. 693—698 (1987).

T. Gengenbach, H. Pagnia, and N. Sotnik
*The Memory Threshold Voltage of MIM Diodes
(73.40 and 85; S10.1)
(a) Vol. **100**, No. 2, p. K199—K204 (1987).
Erratum
(a) Vol. **102**, No. 2, p. K141 (1987).

Yu. A. Firsov, E. P. Nakhmedov, V. N. Prigodin, and W. Weller
Corrections from Electron–Electron Interaction to the One-Particle Density of States in Strongly Anisotropic Two-Dimensional Systems
(71.20; 73.40)
(b) Vol. **141**, No. 1, p. 111—118 (1987).

L. Wendler and R. Pechstedt
Dynamical Screening, Collective Excitations, and Electron–Phonon Interaction in Heterostructures and Semiconductor Quantum Wells. Application to Double Heterostructures
(71.38 and 71.45; 68.30; 73.40; 77.30; 78.65; S7.12; S7.15)
(b) Vol. **141**, No. 1, p. 129—150 (1987).

Yu. B. Gaididei and I. I. Yakimenko
On Trigger Properties of Molecular Organizates. II. Switch Waves in Donor–Acceptor Systems with Strong Electron–Deformation Interaction
(72.10; 73.40)
(b) Vol. **141**, No. 1, p. 213—220 (1987).

P. Enders
Photon Recycling in Double Heterostructures. III. Variational Calculations
(78.55; 72.90; 73.40)
(b) Vol. **141**, No. 1, p. 317—323 (1987).

R. P. Singh and S. L. Singh
*Photoelectrochemical Solar Cells Using Vacuum-Evaporated Zinc Phosphide Films
(73.40; 72.40; S7)
(a) Vol. **101**, No. 1, p. 129—135 (1987).

L. Šamaj
*The Lifetime of Minority Carriers in Polycrystalline Semiconductors
(72.20 and 73.40; S5.11)
(a) Vol. **101**, No. 1, p. 137—141 (1987).

I. Thurzo, J. Bartoš, and E. Pinčík
*On the Structure of the GaAs–Plasma Oxide Interface
(73.40; S7.12)
(a) Vol. **101**, No. 1, p. 165—171 (1987).

B. Bouchikhi, C. Michel, S. Ravelet, and B. Lepley
*Preparation and Electrical Properties of Thin Native Oxide Double-Layer Insulator Films on n-Type InP
(68.55 and 73.40; S7.11)
(a) Vol. **101**, No. 1, p. 173—184 (1987).

DEYAN HĚ, FANGQING ZHANG, JINGZHONG CUI, and GUANGHUA CHEN
*Interface Defects in Sputtered a-Si:H/a-C:H Multilayers
(73.40; 76.30; 79.20; S5; S5.11)
(b) Vol. **142**, No. 1, p. K35—K37 (1987).

L. WENDLER and M. I. KAGANOV
*Helicon Waves in Semiconductor Superlattices under Quantum Hall Effect Conditions
(73.40; S7.12; S7.15)
(b) Vol. **142**, No. 1, p. K63—K68 (1987).

I. G. KIRNAS, P. G. LITOVCHENKO, E. E. PETROSYAN, and N. N. PASHCHUK
*Calculation of Bulk Charge and Electric Field Profiles in One-Open-Face Coaxial γ-Detectors Using Experimental *C–U* Characteristics
(73.40; S5.12)
(a) Vol. **102**, No. 1, p. 315—319 (1987).

G. F. CEROFOLINI
A Model for the Non-Equilibrium, Pure-Generation Centre in Silicon
(71.55; 73.40; S5.11)
(a) Vol. **102**, No. 1, p. 345—348 (1987).

G. GOBSCH, D. SCHULZE, and G. PAASCH
*Inversion Layers in InSb Bicrystals with a Charged Grain Boundary of Finite Thickness
(71.25 and 73.40; S7.13)
(b) Vol. **142**, No. 2, p. K119—K124 (1987).

S. S. KUBAKADDI, B. G. MULIMANI, and N. S. SANKESHWAR
*On the Electrical Conductivity and Thermopower in Quasi-Two-Dimensional Semiconductor Quantum Well Structures in Quantizing Magnetic Field
(73.40)
(b) Vol. **142**, No. 2, p. K131—K135 (1987).

G. J. PAPAIOANNOU
Electron Radiation Effects on GaAs MESFETs
(61.80 and 85; 73.40; S7.12)
(a) Vol. **102**, No. 2, p. 843—848 (1987).

K. KIBICKAS, J. PARŠELIŪNAS, and Š. VASILIAUSKAS
*Transient Hot Electron Phenomena in GaAs n^+–n–n^+ Structures at 300 K
(72.20 and 73.40; S7.12)
(a) Vol. **102**, No. 2, p. K99—K102 (1987).

D. I. TSIULYANU, E. P. KOLOMEYKO, and V. N. STAMOV
*Photoelectric Phenomena in Isotope Vitreous As_2S_3–Silicon Single-Crystal Heterojunctions
(72.40 and 73.40; S5.11; S8)
(a) Vol. **102**, No. 2, p. K103—K107 (1987).

CH. SCHNITTLER and S. SCHERF
*Alternative Quasi-Classical Transport Equations. II. Boltzmann's Equation and the Electron Distribution Function for MIS Systems
(73.40; 73.25; S5.11)
(b) Vol. **143**, No. 1, p. 325—333 (1987).

S. SCHERF, CH. SCHNITTLER, and G. PAASCH
*Alternative Quasi-Classical Transport Equations. III. Charge Transport in MIS Systems with Inclusion of Diffuse Surface Scattering and Quantum Size Effects
(73.40; 73.25; S5.11)
(b) Vol. **143**, No. 2, p. 795—805 (1987).

R. OPILSKI and M. URBANCZYK
*Measurement of Free-Carrier Distributions by Means of Nondestructive SAW-TAV Spectroscopy
(73.40; S5.11)
(a) Vol. **103**, No. 2, p. 637—642 (1987).

G. F. CEROFOLINI, M. L. POLIGNANO, H. BENDER, and C. CLAEYS
*The Role of Dopant and Segregation Annealing in Silicon p–n Junction Gettering
(61.70 and 73.40; S5.11)
(a) Vol. **103**, No. 2, p. 643—654 (1987).

A. POPOV, S. ZHENEVA, I. IVANOV, L. PAPADIMITRIOU, and C. A. DIMITRIADIS
*Deep Levels in Polycrystalline Silicon Solar Cells
(71.55 and 73.40; S5.11)
(a) Vol. **103**, No. 2, p. K99—K102 (1987).

M. SUHRKE, R. KEIPER, and O. ZIEP
*Temperature Dependence of the Mobility in Heterostructures Modified by Multi-Subband Effects
(73.40)
(b) Vol. **144**, No. 1, p. K7—K12 (1987).

C. TRALLERO GINER and F. COMAS
*Electron–LO-Phonon Interaction in Semiconductor Quantum-Well-Heterostructures
(71.38 and 73.40)
(b) Vol. **144**, No. 1, p. K19—K22 (1987).

W. LORENZ
*Chemical Bond Formation Processes on Semiconductor Interfaces at Electronic Equilibrium or Non-equilibrium
(68.45 and 73.40; 71.20; 73.30)
(b) Vol. **144**, No. 2, p. 585—599 (1987).

YI-BEN XIA, E. BANGERT, and G. LANDWEHR
*Low-Temperature Magnetoresistance in Medium-Angle Grain Boundaries in Germanium
(73.40; S5.12)
(b) Vol. **144**, No. 2, p. 601—608 (1987).

V. M. JALI, S. S. KUBAKADDI, and B. G. MULIMANI
Quantum Theory of Thermopower in Quantum Well Wires
(72.20; 73.40)
(b) Vol. **144**, No. 2, p. 739—744 (1987).

73.60. Electronic properties of thin films

J. BULLOT and M. P. SCHMIDT
Physics of Amorphous Silicon–Carbon Alloys **(Review Article)**
(61.40; 71.20; 71.55; 72.40; 73.60; 78.20; 78.55; 78.60; S6)
(b) Vol. **143**, No. 2, p. 345—418 (1987).

H. RÜBEL, B. SCHRÖDER, W. FUHS, J. KRAUSKOPF, T. RUPP, and K. BETHGE
*IR Spectroscopy and Structure of RF Magnetron Sputtered a-SiC:H Films
(68.55 and 73.60; 78.30; S6)
(b) Vol. **139**, No. 1, p. 131—143 (1987).

K. ZAKRZEWSKA, T. PISARKIEWICZ, and A. CZAPLA
*Scattering of Charge Carriers in Reactively Sputtered $CdIn_2O_4$ Thin Films
(73.60; 78.65; S10; S11.2)
(a) Vol. **99**, No. 1, p. 141—149 (1987).

ZHANG FANGQING, XU, XIXIANG, and CHEN GUANGHUA
*Persistent Photoconductivity in a-Si:H/a-$Si_{1-x}C_x$:H Multilayers
(73.60; 72.40; S5.11; S6)
(a) Vol. **99**, No. 1, p. 165—170 (1987).

F. COMAS, C. TRALLERO GINER, and J. TUTOR
*Ohmic Mobility of Quasi-One-Dimensional Semiconducting Structures. Acoustic and Piezoelectric Phonon Scattering
(72.20 and 73.60; S7.12; S8.11)
(b) Vol. **139**, No. 2, p. 433—440 (1987).

M. ZOR and C. A. HOGARTH
*High-Field Electrical Conduction in Thin Films of Polyethylene
(73.60; S12)
(a) Vol. **99**, No. 2, p. 513—519 (1987).

Y. OCHIAI
*Dose Dependence of $N(E_F)$ in Evaporated a-Si
(73.60; 61.70; S5.11)
(a) Vol. **99**, No. 2, p. K87—K90 (1987).

C. MAJUMDAR, A. B. MAITY, and A. N. CHAKRAVARTI
*Effect of Size Quantization on Photoemission from Ultrathin Films of Wide-Gap Semiconductors
(73.60 and 79.60; S7.12)
(b) Vol. **140**, No. 1, p. K7—K12 (1987).

R. M. CÂNDEA, R. TURCU, G. BORODI, and I. BRATU
*Effects of Thermal Annealing in Air on VE, COD, and CAD PbSe Films
(68.55 and 73.60; 78.65; S8)
(a) Vol. **100**, No. 1, p. 149—155 (1987).

S. ORZESZKO, K. FABISIAK, and F. ROZPŁOCH
*Optical and Mobility (Electrical) Gaps of Arc Evaporated Hydrogenated Amorphous Carbon Films
(73.60 and 78.65; S5)
(a) Vol. **101**, No. 1, p. 159—163 (1987).

A. J. DANN, M. R. FAHY, C. JEYNES and M. R. WILLIS
*Electrical Properties of Ion Implanted PcAlF Films
(73.60; S12.1)
(a) Vol. **101**, No. 1, p. K45—K48 (1987).

C. MAJUMDAR, M. K. BOSE, A. B. MAITY, and A. N. CHAKRAVARTI
*Effect of Size Quantization on Field Emission from Ultrathin Films of Degenerate Wide-Gap Semiconductors
(73.60; 79.70; S7.12)
(b) Vol. **141**, No. 2, p. 435—439 (1987).

A. ZEGHIB, J. LE BAS, and C. VAUTIER
*Electrical and Magnetic Properties of Amorphous Ni_xAg_{1-x} Thin Films
(73.60 and 75.70; S1.2; S1.3)
(a) Vol. **101**, No. 2, p. 487—493 (1987).

EL SAYED MEHANNA, S. ARAJS, and H. F. HELBIG
*Effects of Stresses on the Néel Temperature of Thin Polycrystalline Chromium Films
(73.60 and 75.70; S1.2)
(a) Vol. **101**, No. 2, p. K129—K133 (1987).

D. T. RASULOV
*Electrical Properties of $Zn_xCd_{1-x}S$ Films
(73.60; S8.15)
(a) Vol. **101**, No. 2, p. K139—K142 (1987).

L. Q. MINH, T. CHOT, N. N. DINH, N. N. XUAN, N. T. BINH, D. M. PHUOC, and N. T. BINH
*A New Organic Photoconductor in the Near Infrared Region of Diode Lasers
(73.60 and 78.65; S12)
(a) Vol. **101**, No. 2, p. K143—K146 (1987).

E. S. M. HIGGY, S. M. YOUSSEF, S. A. EL-SAHHAR, and H. M. TALAAT
*Electrical Properties of Thin Polycrystalline Lead Films
(73.60; S4)
(a) Vol. **102**, No. 1, p. 301—306 (1987).

N. C. HALDER
*Tunneling and Hopping Parameters in Ta-Oxide Thick Films
(73.60; S10.1)
(a) Vol. **102**, No. 1, p. 307—313 (1987).

J. P. GUESDON, B. KOBBI, C. JULIEN, and M. BALKANSKI
*Electrical and Photovoltaic Properties of In_xSe_{1-x} Thin Films
(72.40 and 73.60; S8)
(a) Vol. **102**, No. 1, p. 327—335 (1987).

K. Ikezaki
*Thermally Stimulated Currents in FEP Films with Sputtered Al Electrodes
(73.60; S12)
(a) Vol. **103**, No. 2, p. 481—489 (1987).

K. Tennakone, A. H. Jayatissa, C. A. N. Fernando, S. Wickramanayake, S. Punchihewa, L. K. Weerasena, and W. D. R. Premasiri
*Semiconducting and Photoelectrochemical Properties of n- and p-Type β-CuCNS
(73.60 and 78.65; 68.55; 72.40; S11)
(a) Vol. **103**, No. 2, p. 491—497 (1987).

C. Majumdar, A. B. Maity, and A. N. Chakravarti
*Quantized Photoemission from Quasi-Two-Dimensional Structures of Wide-Gap Semiconductors in a Strong Magnetic Field
(73.60 and 79.60; S7.12)
(b) Vol. **144**, No. 1, p. K13—K18 (1987).

A. Möbius
The Metal–Semiconductor Transition in Amorphous $Si_{1-x}Cr_x$ Films. Detection of the Transition by Room-Temperature Measurements
(71.30 and 72.60; 72.20; 73.60; S1.61)
(b) Vol. **144**, No. 2, p. 759—766 (1987).

M. Zor and C. A. Hogarth
*Some DC and AC Properties of Copper-Loaded Polypropylene Films
(73.60; S12)
(a) Vol. **104**, No. 2, p. 761—768 (1987).

73.90. Other topics in electrical properties of surfaces, interfaces, and thin films

D. Lovrić and B. Gumhalter
Physisorption and Chemisorption Effects on Atom-Adsorbate van der Waals Potentials. Application to He and He Scattering from Adsorbed CO
(68.20 and 73.90; 79.20)
(b) Vol. **139**, No. 2, p. 423—432 (1987).

F. Berz and J. A. Morice
Accuracy of the Transmission Coefficient across Parabolic Barriers as Obtained from a Generalized WKB Approach
(72.90; 73.90)
(b) Vol. **139**, No. 2, p. 573—582 (1987).

74. Superconductivity

74.10. Occurrence, critical temperature

S. L. KAKANI and U. N. UPADHYÀY
*Coexistence of Superconductivity and Ferromagnetism in Rare Earth Ternary Systems **(Review Article)**
(74.10 and 75.30; S1.4)
(a) Vol. **99**, No. 1, p. 15—36 (1987).

V. E. ANTONOV, I. T. BELASH, O. V. ZHARIKOV, and A. V. PALNICHENKO
*A Study on Superconductivities of High-Pressure Phases in the Re–H, Ru–H, Rh–H, and Ni–H Systems
(74.10; S1)
(b) Vol. **142**, No. 2, p. K155—K159 (1987).

I. M. CHAPNIK
*Application of Dimensional Method to Superconductivity
(74.10 and 74.30)
(b) Vol. **144**, No. 1, p. 225—232 (1987).

B. YA. SHAPIRO and L. V. EFIMOVA
*The Surface Superconductivity of Layered Metals
(74.10)
(b) Vol. **144**, No. 1, p. 437—448 (1987).

J. KAMARÁD, Z. ARNOLD, and E. POLLERT
*Pressure Dependence of the Critical Temperature of Superconductivity in $YBa_2Cu_3O_{7-x}$
(74.10; S10.15)
(b) Vol. **144**, No. 1, p. K39—K43 (1987).

G. J. HYLAND
*Mott-Insulation, High Frequencies, and the High Transition Temperatures of the New Ceramic Superconductors
(74.10)
(b) Vol. **144**, No. 2, p. 753—758 (1987).

S. L. DRECHSLER and N. M. PLAKIDA
*The Isotope Effect within the Anharmonic Model for High-T_c Superconductors
(74.10; S10.15)
(b) Vol. **144**, No. 2, p. K113—K117 (1987).

74.20. Theory

Z. BAK and M. PIASECKI
*Phenomenological Theory of Weak Ferromagnetic Superconductors
(74.20 and 75.30)
(b) Vol. **140**, No. 2, p. K149—K153 (1987).

J. Monecke
*Material Constants of Composites — The Permeability of Mixtures of Superconducting with Non-Superconducting Phases
(74.20; 74.30)
(b) Vol. **143**, No. 1, p. K43—K45 (1987).

L. S. Brizhik and A. S. Davydov
*Soliton Mechanism of Superconductivity in Organic Quasi-One-Dimensional Crystals
(74.20; 71.45; S12.1)
(b) Vol. **143**, No. 2, p. 689—698 (1987).

74.30. General properties

S. De Stefano, A. Di Chiara, G. Peluso, and U. Scotti di Uccio
*Aspects of Magnetic Penetration in Vanadium Based Josephson Junctions
(74.30 and 74.50; S1.2)
(a) Vol. **99**, No. 2, p. K109—K110 (1987).

D. P. Hampshire and H. Jones
*The Detailed Investigation of the Critical Current Density of Optimized Nb_3Sn Modified Jelly Roll Wire throughout Its Entire Superconducting Phase Including a Comparative Study with a Similar Non-Optimized Composite
(74.30)
(b) Vol. **141**, No. 1, p. 229—237 (1987).

A. M. Gabovich, A. S. Gerber, and A. S. Shpigel
*Thermodynamics of Superconductors with Charge- and Spin-Density Waves. Δ/T_c Ratio and Paramagnetic Limit
(74.30)
(b) Vol. **141**, No. 2, p. 575—587 (1987).

N. B. Ivanov and T. M. Mishonov
*Heat Capacity of a Thick Superconducting Plate with Proximity Effect
(74.30)
(b) Vol. **142**, No. 1, p. K49—K52 (1987).

J. Monecke
Material Constants of Composites — The Permeability of Mixtures of Superconducting with Non-Superconducting Phases
(74.20; 74.30)
(b) Vol. **143**, No. 1, p. K43—K45 (1987).

I. M. Chapnik
*Application of Dimensional Method to Superconductivity
(74.10 and 74.30)
(b) Vol. **144**, No. 1, p. 225—232 (1987).

S. Takács
*Possible Restriction of the Critical Current Density in High Temperature Superconductors
(74.30)
(b) Vol. **144**, No. 2, p. K125—K128 (1987).

74.40. Fluctuations and critical effects

A. V. Gabrelyan, Yu. G. Morozov, and E. A. Chernov
Effect of Magnetic Field on Superconductivity of Fine-Dispersed Systems
(74.70; 74.40; S4)
(a) Vol. **101**, No. 2, p. 521—530 (1987).

74.50. Tunneling phenomena, Josephson effect, and proximity effects

S. De Stefano, A. Di Chiara, G. Peluso, and U. Scotti di Uccio
*Aspects of Magnetic Penetration in Vanadium Based Josephson Junctions
(74.30 and 74.50; S1.2)
(a) Vol. **99**, No. 2, p. K109—K110 (1987).

74.60. Type-II superconductivity

E. H. Brandt and U. Essmann
*The Flux-Line Lattice in Type-II Superconductors **(Review Article)**
(74.60)
(b) Vol. **144**, No. 1, p. 13—38 (1987).

A. G. Saif and L. N. Shehata
*Magnetic Pinning Interaction of a Type-II Superconducting Cylinder
(74.60)
(b) Vol. **139**, No. 2, p. 547—557 (1987).

P. Tekiel
*Mixed State of Hollow Current-Carrying Type II Superconductor
(74.60)
(b) Vol. **140**, No. 2, p. K133—K136 (1987).

A. G. Saif
*AC Losses in a Type-II Superconducting Slab Including Surface Barrier Effects
(74.60)
(b) Vol. **141**, No. 1, p. 203—211 (1987).

74.70. Superconducting materials

FARID A. KHWAJA and F. BROUERS
*Short-Range Order and Superconductivity in Binary Alloys Containing V, Nb, and Ta
(74.70; 61.55; S1; S1.2)
(b) Vol. **139**, No. 2, p. 565—571 (1987).

W. ZACHARKO
*Experimental Investigations on the Anisotropy of the Upper Critical Field in Hexagonal Intrinsic Type II Superconductor $InSn_4$
(74.70; S4)
(b) Vol. **140**, No. 1, p. K33—K36 (1987).

B. FELLMUTH, D. ELEFANT, and J.-I. MÖNCH
*Investigation of the Superconducting Transition of Ultra-High-Purity Niobium
(74.70; S1)
(a) Vol. **100**, No. 2, p. 597—605 (1987).

A. V. GABRELYAN, YU. G. MOROZOV, and E. A. CHERNOV
*Effect of Magnetic Field on Superconductivity of Fine-Dispersed Systems
(74.70; 74.40; S4)
(a) Vol. **101**, No. 2, p. 521—530 (1987).

R. HERRMANN, N. KUBICKI, TH. SCHURIG, H. DWELK, U. PREPPERNAU, A. KRAPF, H. KRÜGER, H.-U. MÜLLER, L. ROTHKIRCH, W. KRAAK, W. BRAUNE, N. PRUSS, G. NACHTWEI, F. LUDWIG, and E. KEMNITZ
*Preparation and Properties of High T_c Superconducting Y–Ba–Cu–O Systems
(74.70; S10.15)
(b) Vol. **142**, No. 1, p. K53—K56 (1987).

C. F. VAN BRUGGEN
*Monovalent-Copper (I; $3d^{10}$, $S = 0$) Induced Oxygen-2p Valence Band Holes with Concomitant Attractive Interaction inside Low-Dimensional Oxygen Networks
(71.25 and 74.70; 75.20; S1.4; S10.15)
(b) Vol. **142**, No. 2, p. 537—550 (1987).

K. ROGACKI, C. SUŁKOWSKI, A. J. ZALESKI, Z. BUKOWSKI, M. CISZEK, R. HORYŃ, E. TROJNAR, and J. KLAMUT
*Critical Current Densities for La–Sr–Cu–O and Y–Ba–Cu–O Samples
(74.70; S10)
(a) Vol. **102**, No. 2, p. K117—K119 (1987).

O. SMRČKOVÁ, D. SÝKOROVÁ, J. DOMINEC, J. JUREK, and L. SMRČKA
*On the Superconductivity in the Y–Ba–Cu–O System
(74.70; S10)
(a) Vol. **103**, No. 1, p. K33—K36 (1987).

75. Magnetic properties and materials

75.10. General theory and models of magnetic ordering

(see also 71.25. Nonlocalized single-particle electronic states, 71.70. Level splitting and interactions)

W. Nolting and A. Ramakanth
*Magnetic Order in Intermediate Valence Systems
(75.10; 75.30; 75.50)
(b) Vol. **139**, No. 1, p. 281—298 (1987).

J. Katriel
*Continued-Fraction Approximation for the Inverse Brillouin Function
(75.10)
(b) Vol. **139**, No. 1, p. 307—310 (1987).

G. A. Farias and L. L. Gonçalves
*Short-Time Behaviour of the Transverse Auto-Correlation Function of the 1D XY-Model
(75.10)
(b) Vol. **139**, No. 1, p. 315—320 (1987).

R. B. Prabhu
*Mix-Valent Europium Based Systems in Interconfiguration Fluctuation Model
(75.10; S1.61; S1.62)
(b) Vol. **139**, No. 1, p. K69—K73 (1987).

T. Iwashita and N. Uryû
*Magnetic Properties of the Ising Ferromagnet of Large Spin with Biquadratic Exchange Interaction and Uniaxial Anisotropy
(75.10; 75.30)
(b) Vol. **139**, No. 2, p. 597—607 (1987).

G. Bulk and W. Nolting
*Many-Body Effects in Antiferromagnetic 4f-Systems
(75.10; 71.25)
(b) Vol. **140**, No. 1, p. 261—271 (1987).

J. Sznajd, G. Kontrym-Sznajd, and A. Drzewinski
*Renormalization Group Approach to Antiferromagnetic Quantum XY Model on a Triangular Lattice
(75.10)
(b) Vol. **140**, No. 2, p. K155—K160 (1987).

U. Lindner
*To Applicability of Hamilton's Equations in Quantum Many-Body Problems
(75.10; 63.10; 71.10)
(b) Vol. **141**, No. 1, p. 263—268 (1987).

G. Fischer and A. Herr
*Representation of Energetical and Low-Field Magnetic Properties of $J = 5/2$ Rare Earth Ion States in Tetragonal Crystal Fields
(75.10; S1.4)
(b) Vol. **141**, No. 2, p. 589—598 (1987).

75.20. Diamagnetism and paramagnetism

I. OKAMOTO, T. MIYAZAKI, and M. TAKAHASHI
*Change of Magnetic Properties with Heat Treatments in Rapidly Quenched $Cu_{99.2}Fe_{0.8}$ Alloys
(75.20 and 75.50; S1.1; S1.2)
(a) Vol. **99**, No. 1, p. 267—274 (1987).

L. A. HUSSAIN and O. AKINLADE
A Model Potential and Its Applications
(71.45 and 72.15; 75.20)
(b) Vol. **139**, No. 2, p. K123—K128 (1987).

T. SATOW, O. UEMURA, K. MATSUMOTO, and S. OKAMURA
Electronic Properties of Liquid Sb–Se and Bi–Se Alloys
(72.15; 61.25; 75.20; S4)
(b) Vol. **140**, No. 1, p. 233—242 (1987).

H. J. SCHELLNHUBER and H. URBSCHAT
Analyticity Breaking of Wave Functions and Fractal Phase Diagram for Simple Incommensurate Systems
(71.10; 75.20)
(b) Vol. **140**, No. 2, p. 509—519 (1987).

M. BUDZYŃSKI, O. I. KOCHETOV, M. SUBOTOWICZ, H. NIEZGODA, H. SPUSTEK, and W. TAŃSKA-KRUPA
*Influence of Hydrogenation on the Hyperfine Interactions in $(Y_{0.97}Hf_{0.03})Fe_2$
(75.20; 75.25; 75.50; 76.80; S1.1; S1.4)
(b) Vol. **140**, No. 2, p. 589—594 (1987).

M. KUCHARCZYK and S. OLSZEWSKI
*Modified Thomas-Fermi Model Applied to Diamagnetic Susceptibilities of Crystal Ions
(75.20)
(b) Vol. **140**, No. 2, p. K141—K144 (1987).

F. M. KAMARLY and A. YU. YANGIROV
*Magnetic Susceptibility of the $(SnTe)_{1-x}$–$(EuTe)_x$ System
(75.20; 72.80; S8.16)
(a) Vol. **101**, No. 1, p. K61—K63 (1987).

C. F. VAN BRUGGEN
Monovalent-Copper (I; $3d^{10}$, $S = 0$) Induced Oxygen-2p Valence Band Holes with Concomitant Attractive Interaction inside Low-Dimensional Oxygen Networks
(71.25 and 74.70; 75.20; S1.4; S10.15)
(b) Vol. **142**, No. 2, p. 537—550 (1987).

75.25. Spin arrangements in magnetically ordered materials
(neutron studies, etc.)

75.30. Magnetically ordered materials, other intrinsic properties
(for critical point effects, see 75.40)

S. L. Kakani and U. N. Upadhyaya
*Coexistence of Superconductivity and Ferromagnetism in Rare Earth Ternary Systems **(Review Article)**
(74.10 and 75.30; S1.4)
(a) Vol. **99**, No. 1, p. 15—36 (1987).

I. A. Ovidko and A. E. Romanov
Topological Excitations (Defects, Solitons, Textures, Frustrations) in Condensed Media **(Review Article)**
(61.70; 64.60; 71.45; 75.30)
(a) Vol. **104**, No. 1, p. 13—45 (1987).

W. Nolting and A. Ramakanth
Magnetic Order in Intermediate Valence Systems
(75.10; 75.30; 75.50)
(b) Vol. **139**, No. 1, p. 281—298 (1987).

L. Cianchi, A. Cuccoli, M. Mancini, and G. Spina
*A Comment on the Role of the Mixed Spin Relaxation Process in $ZnCO_3 : Fe^{2+}$
(75.30; S11.1)
(b) Vol. **139**, No. 1, p. K53—K56 (1987).

I. Škorvánek and A. Zentko
Effects of Neutron Irradiation on Magnetic Properties of $Fe_{80}B_{20}$ Amorphous Alloys
(75.50; 61.80; 75.30; 75.60; S1.1)
(a) Vol. **99**, No. 1, p. 275—278 (1987).

D. Neogy and T. Purohit
The Behavior of Active Centers in a Laser Host. A Crystal Field Investigation of Sm^{3+} in LaF_3 Single Crystal
(71.70; 65.40; 75.30; S9)
(b) Vol. **139**, No. 2, p. 519—525 (1987).

T. Iwashita and N. Uryû
·Magnetic Properties of the Ising Ferromagnet of Large Spin with Biquadratic Exchange Interaction and Uniaxial Anisotropy
(75.10; 75.30)
(b) Vol. **139**, No. 2, p. 597—607 (1987).

Yu. M. Seidov and A. M. Suleimanov
To the Theory of Two-Magnon Absorption of Light by Surface Collective Excitations in the Singlet–Triplet Ferromagnets
(71.70; 75.30; 78.20; S1.4)
(b) Vol. **140**, No. 1, p. 149—154 (1987).

H. A. Brown
*On a Ferromagnet with a Three-Spin Interaction
(75.30)
(b) Vol. **140**, No. 1, p. K55—K59 (1987).

Z. FRAIT, D. FRAITOVÁ, N. ZÁRUBOVÁ, and P. DUHAJ
Magnetic Surface Anisotropy and Ferromagnetic Resonance in a Stress Annealed FeNiB Amorphous Alloy
(76.50; 75.30; S1.1; S1.2)
(a) Vol. **101**, No. 1, p. 239—245 (1987).

J. KLAMUT and W. PRYSTASZ
Quantum Tunneling at First-Order Phase Transitions in a Uniaxial Antiferromagnet with External Magnetic Field
(64.60; 75.30)
(b) Vol. **142**, No. 1, p. 83—87 (1987).

V. V. VALKOV, T. A. VALKOVA, and S. G. OVCHINNIKOV
*Spectrum of Excitations of an Easy-Plane Ferromagnet ($S = 3/2$) in a Magnetic Field
(75.30; 75.50)
(b) Vol. **142**, No. 1, p. 255—263 (1987).

A. K. RASTOGI and E. P. WOHLFARTH
*Magnetic Field-Induced Transitions in the Mo_4 Cluster Compounds $GaMo_4S_8$ and $GaMo_4Se_8$ Showing Heavy Fermion Behaviour
(75.30; S11)
(b) Vol. **142**, No. 2, p. 569—573 (1987).

M. KURISU, H. KADOMATSU, and H. FUJIWARA
*Magnetic Field Dependences of Two Kinds of Structural Transition Temperatures of $CeAg_{1-x}In_x$
(64.60 and 75.30; 72.15; S1.3; S1.4)
(b) Vol. **142**, No. 2, p. 581—585 (1987).

A. R. HARUTYUNYAN, L. S. GRIGORYAN, and E. G. SHAROYAN
*Magnetic Properties of Free-Base Phthalocyanine Doped with Sodium
(75.30; S12.1)
(b) Vol. **142**, No. 2, p. K169—K172 (1987).

C. AMMARGUELLAT, M. ESCORNE, A. MAUGER, E. BEAUREPAIRE, M. F. RAVET, G. KRILL, F. LAPIERRE, P. HAEN, and C. GODART
Mixed Valence Properties of CeM_2Si_2 (M = Mn, Fe, Co, Ni, Cu)
(71.25; 72.15; 75.30; S1.61)
(b) Vol. **143**, No. 1, p. 159—166 (1987).

S. MANHAS, K. C. KHULBE, D. J. S. BECKETT, G. LAMARCHE, and J. C. WOOLLEY
*Lattice Parameters, Energy Gaps, and Magnetic Properties of the $Cd_xHg_yMn_zTe$ Alloy System
(75.30; 61.60; 71.25; S8.16)
(b) Vol. **143**, No. 1, p. 267—274 (1987).

J. M. DANCE, J. P. CHAMINADE, J. M. MOUTOU, and M. POUCHARD
*Spectroscopic and Magnetic Characterization of the Copper–Tungsten Oxyfluoride $CuWO_3F_2$
(75.30 and 76.30; 78.20; S9.16)
(b) Vol. **143**, No. 1, p. K57—K61 (1987).

J. IÑIGUEZ, J. RIVAS, and F. GUITIÁN
*Thermomagnetic Analysis and Reaction Kinetics of the Sintering of Magnetite Samples
(75.30; S11.2)
(a) Vol. **103**, No. 1, p. K45—K49 (1987).

75.40. Critical-point effects, specific heats, short-range order
(see also 65.40. Heat capacities)

D. HAJDUKOVIĆ and M. MIJATOVIĆ
*One-Dimensional Ising Bridge Model
(75.40)
(b) Vol. **139**, No. 1, p. K47—K51 (1987).

D. SCHULZ and J. SCHREIBER
*On the Specific Heat of a Heisenberg Spin Glass: The Effect of Dynamics
(75.40)
(b) Vol. **139**, No. 1, p. K57—K61 (1987).

M. M. AMADO, J. D. MONTENEGRO, J. M. MOREIRA, M. E. BRAGA, and J. B. SOUSA
*Study of the Magnetic Transitions in $CeFe_2$ with High Resolution Transport Property Measurements
(75.40; 72.15; 75.50; S1.1; S1.4)
(a) Vol. **99**, No. 2, p. 625—630 (1987).

D. V. KAPOR, M. J. ŠKRINJAR, and J. P. ŠETRAJČIĆ
*Specific Heat of an Anisotropic Heisenberg Model
(75.40)
(b) Vol. **140**, No. 1, p. K61—K64 (1987).

M. FÄHNLE and M. HAUG
Phase Transitions in Ordered and Disordered Ferrimagnets. I. Correlated Molecular Field Theory and Monte Carlo Simulations
(75.30; 75.40; 75.50)
(b) Vol. **140**, No. 2, p. 569—579 (1987).

T. TARNÓCZI and A. LOVAS
*Kinetics of Relaxation in Metallic Glasses with and without Cr
(75.40; S1.2)
(a) Vol. **100**, No. 2, p. K177—K179 (1987).

T. HOLEY and M. FÄHNLE
Correlated Molecular Field Theory for the Ferromagnetic Regime of Amorphous Systems
(75.30; 75.40; 75.50)
(b) Vol. **141**, No. 1, p. 253—262 (1987).

S. KRZEMIŃSKI
*Note on the *n*-Vector Glasser Model
(75.40)
(b) Vol. **141**, No. 2, p. K117—K120 (1987).

75.50. Studies of specific magnetic materials

W. NOLTING and A. RAMAKANTH
Magnetic Order in Intermediate Valence Systems
(75.10; 75.30; 75.50)
(b) Vol. **139**, No. 1, p. 281—298 (1987).

V. S. LUTOVINOV
The Long-Wavelength Magnon Relaxation in Ferromagnetic Metals at $T \lessapprox T_c$
(76.50; 75.50)
(b) Vol. **139**, No. 1, p. 299—305 (1987).

I. OKAMOTO, T. MIYAZAKI, and M. TAKAHASHI
*Change of Magnetic Properties with Heat Treatments in Rapidly Quenched $Cu_{99.2}Fe_{0.8}$ Alloys
(75.20 and 75.50; S1.1; S1.2)
(a) Vol. **99**, No. 1, p. 267—274 (1987).

I. ŠKORVÁNEK and A. ZENTKO
*Effects of Neutron Irradiation on Magnetic Properties of $Fe_{80}B_{20}$ Amorphous Alloys
(75.50; 61.80; 75.30; 75.60; S1.1)
(a) Vol. **99**, No. 1, p. 275—278 (1987).

H. GÄRTNER and Z. FRAIT
*The Influence of Ion Plating of Nickel on Amorphous Magnetic Alloy as Detected by Ferromagnetic and Antiferromagnetic Resonance
(75.50 and 76.50; S1.2)
(a) Vol. **99**, No. 1, p. K57—K59 (1987).

K. TERAO and M. SHIMIZU
*Electronic Structure and Magnetism of AFe_2 Compounds with C14 and C15 Laves Structures
(71.25 and 75.50; 75.80; S1.1)
(b) Vol. **139**, No. 2, p. 485—497 (1987).

T. MIYAZAKI, H. TAKADA, and M. TAKAHASHI
*Magnetic Properties of Rapidly Quenched $Fe_{75}B_{25-x}Nd_x$ $(0 \leqq x \leqq 25)$ Alloys
(75.50; 75.60; S1.1)
(a) Vol. **99**, No. 2, p. 611—618 (1987).

M. M. AMADO, J. D. MONTENEGRO, J. M. MOREIRA, M. E. BRAGA, and J. B. SOUSA
Study of the Magnetic Transitions in $CeFe_2$ with High Resolution Transport Property Measurements
(75.40; 72.15; 75.50; S1.1; S1.4)
(a) Vol. **99**, No. 2, p. 625—630 (1987).

K. TABATA and H. KIDO
*Preparation and Magnetic Properties of $La_{1-x}Th_xCoO_3$ Compounds
(75.50; S1.4)
(a) Vol. **99**, No. 2, p. K121—K123 (1987).

M. Wurlitzer
*Manifestations of Induced Magnetic Anisotropy in Zr-Doped Yttrium Iron Garnet
(75.50; S11.2)
(a) Vol. **100**, No. 2, p. 607—620 (1987).

H. Schäfer and H. H. Mende
*Stray Field Profile of Grain-Oriented Rectangular 3% SiFe Sheets in Dependence on Vertical and In-Plane Magnetic Fields
(75.50; S1.1)
(a) Vol. **100**, No. 2, p. 621—632 (1987).

M. Jurczyk
*Crystallographic and Magnetic Characteristics of $Ce_3Co_{20}B$ Compound
(61.60 and 75.50; S11)
(a) Vol. **100**, No. 2, p. K173—K176 (1987).

T. Holey and M. Fähnle
Correlated Molecular Field Theory for the Ferromagnetic Regime of Amorphous Systems
(75.30; 75.40; 75.50)
(b) Vol. **141**, No. 1, p. 253—262 (1987).

A. Lehmeyer and L. Merten
*Shape of Magnetic Reststrahlen-Bands Caused by Magnon-Polaritons in Uniaxial Ferrites
(71.36 and 75.50; 78.30)
(b) Vol. **141**, No. 1, p. 269—286 (1987).

M. Jurczyk, A. Kowalczyk, and A. Wrzeciono
*Magnetic Properties of $Nd_2Fe_{14-x}Si_xB$ Compounds
(75.50; S1.1)
(a) Vol. **101**, No. 1, p. K65—K68 (1987).

L. N. Bezmaternykh, N. A. Sokolova, and V. L. Temerov
*Magnetoelastic and Magnetostatic Waves in Iron Garnets with Dislocation Accumulation
(75.50; S11.2)
(a) Vol. **101**, No. 2, p. 543—548 (1987).

D. Eckert, D. Hinz, A. Handstein, and J. Schneider
*Thermal Ageing and Coercivity of Sintered Nd–Fe–B Magnets
(75.50; 75.40; 75.60; S1.1; S1.4)
(a) Vol. **101**, No. 2, p. 563—566 (1987).

B. V. Costa and A. S. T. Pires
*Properties of Kinks in a Driven Damped Anisotropic Magnetic Chain
(71.45 and 75.50; S12.1)
(b) Vol. **142**, No. 1, p. 117—123 (1987).

H. Matsui, S. Suzuki, and M. Tamaki
*Characteristic Temperatures in the Electrical Resistivity and Magnetization of β-U_2N_3
(72.15 and 75.50; S1.62)
(a) Vol. **103**, No. 1, p. 199—203 (1987).

A. Hauet, J. Teillet, B. Hannoyer, and M. Lenglet
Mössbauer Study of Co and Ni Ferrichromites
(76.80; 75.50; S11.2)
(a) Vol. **103**, No. 1, p. 257—261 (1987).

F. J. Kedves, M. Hordós, L. Potocký, É. Kisdi-Koszó, and J. Kováč
*Some Structural Aspects of Magnetic Properties and Electrical Resistance of Amorphous FeCrB Alloys
(72.15 and 75.50; S1.1)
(a) Vol. **103**, No. 1, p. 273—280 (1987).

J. Sosnowski
Superconducting and Magnetic Properties of A15-Type Nb_3Ga Alloys Doped with 3d Impurities
(74.70; 75.50; S1)
(a) Vol. **103**, No. 2, p. 531—542 (1987).

A. I. C. Persiano and R. D. Rawlings
*A Mössbauer Investigation of Equiatomic FeCo with Vanadium and Niobium Additions
(75.50 and 76.80; S1.1; S1.2)
(a) Vol. **103**, No. 2, p. 547—556 (1987).

H. Bosse and H. Gärtner
*A Systematic Ferromagnetic Resonance Study of Fe-, Ni-, and Cr-Ion Plated Amorphous Ribbons
(75.50 and 76.50; S1.1; S1.2)
(a) Vol. **103**, No. 2, p. 565—570 (1987).

A. G. Rustamov, G. I. Gashimov, and A. A. Mustafaev
*Investigation of the Electrophysical Properties of an NiS Single Crystal in the Magnetic-Transformation Region
(72.15 and 75.50; S1.63)
(a) Vol. **103**, No. 2, p. K131—K135 (1987).

H. Kronmüller
*Theory of Nucleation Fields in Inhomogeneous Ferromagnets
(75.50 and 75.60)
(b) Vol. **144**, No. 1, p. 385—396 (1987).

W.-U. Kellner, M. Fähnle, H. Kronmüller, and S. N. Kaul
The Magnetic Phase Transition in Amorphous and Disordered Crystalline Ferromagnets
(75.30; 75.40; 75.50; S1.1; S1.2)
(b) Vol. **144**, No. 1, p. 397—409 (1987).

75.60. Domain effects, magnetization curves, and hysteresis

A. ROJEK and C. SUŁKOWSKI
*Anomalous Magnetic and Electrical Properties of the $EuRh_{1.5}Sn_{4.2}$ Compound
(72.15 and 75.60; S1.4)
(b) Vol. **139**, No. 1, p. K63—K67 (1987).

I. ŠKORVÁNEK and A. ZENTKO
Effects of Neutron Irradiation on Magnetic Properties of $Fe_{80}B_{20}$ Amorphous Alloys
(75.50; 61.80; 75.30; 75.60; S1.1)
(a) Vol. **99**, No. 1, p. 275—278 (1987).

K.-H. MÜLLER, A. HANDSTEIN, D. ECKERT, and J. SCHNEIDER
*The Dip in Magnetization Curves of Sintered Nd–Fe–B Permanent Magnets
(75.60; S1.1; S1.4)
(a) Vol. **99**, No. 1, p. K61—K64 (1987).

T. MIYAZAKI, H. TAKADA, and M. TAKAHASHI
Magnetic Properties of Rapidly Quenched $Fe_{75}B_{25-x}Nd_x$ ($0 \leqq x \leqq 25$) Alloys
(75.50; 75.60; S1.1)
(a) Vol. **99**, No. 2, p. 611—618 (1987).

TH. MOCH, B. LIPPOLD, and A. HANDSTEIN
*Erratum to: Spin Glass Like Behaviour of a Sputtered $Cu_{94.6}Tb_{5.4}$ Alloy
(75.60; S1.2)
(a) Vol. **100**, No. 2, p. K205 (1987).

R. KRISHNAN and M. DANCYGIER
*High Field Susceptibility Peak in Some NiFeSiB Amorphous Alloys
(75.60; S1.1; S1.2)
(b) Vol. **141**, No. 1, p. K71—K74 (1987).

N. AMIN, S. ARAJS, and E. MATIJEVIĆ
*Magnetic Properties of Uniform Spherical Magnetite Particles Prepared from Ferrous Hydroxide Gels
(75.60; S10)
(a) Vol. **101**, No. 1, p. 233—238 (1987).

J. F. FUERTES and M. TEJEDOR
*Pinning Effect on the Domain Structures of an Imperfect Iron Whisker
(75.60; S1.1)
(a) Vol. **101**, No. 2, p. 549—554 (1987).

D. ECKERT, D. HINZ, A. HANDSTEIN, and J. SCHNEIDER
Thermal Ageing and Coercivity of Sintered Nd–Fe–B Magnets
(75.50; 75.40; 75.60; S1.1; S1.4)
(a) Vol. **101**, No. 2, p. 563—566 (1987).

R. Berthe, A. Birkner, and U. Hartmann
*Current-Induced Changes of the Magnetic Domain Structure of Iron Whiskers
(75.60; S1.1)
(a) Vol. **103**, No. 2, p. 557—564 (1987).

J. Šebek, J. Stehno, V. Nekvasil, S. Šafrate, J. Šrámek, J. Hruška, L. Havela, V. Sechovský, P. Svoboda, and V. Valvoda
*Magnetic Behavior of $REBa_2Cu_3O_7$ (RE = Y, Gd, Dy, Ho)
(75.60; S10)
(a) Vol. **103**, No. 2, p. K119—K123 (1987).

H. J. Blythe, J. Rivas, and J. Iñiguez
*An Investigation of the Magnetic After-Effect in Magnetite at High Temperatures
(75.60; S10.1)
(a) Vol. **103**, No. 2, p. K137—K140 (1987).

H. Kronmüller
*Theory of Nucleation Fields in Inhomogeneous Ferromagnets
(75.50 and 75.60)
(b) Vol. **144**, No. 1, p. 385—396 (1987).

F. Walz and H. J. Blythe
Binding Enthalpies of Carbon–Vacancy Complexes in α-Fe and Their Compatibility with Different Annealing Models
(61.70; 61.80; 71.55; 75.60; S1.1)
(a) Vol. **104**, No. 1, p. 343—356 (1987).

A. Hoffmann and H. Kronmüller
Diffusion of Hydrogen Isotopes in Amorphous Ferromagnets. I. Short-Range Processes Studied by Magnetic Aftereffect
(66.30; 61.40; 75.60; S1.1)
(a) Vol. **104**, No. 1, p. 381—396 (1987).

J. Pastushenkov, K.-D. Durst, and H. Kronmüller
*Domain Observations under Applied Fields of Sintered $Fe_{77}Nd_{15}B_8$ Permanent Magnets
(75.60; S1.1)
(a) Vol. **104**, No. 1, p. 487—495 (1987).

A. Hofmann and H. Kronmüller
Diffusion of Hydrogen Isotopes in Amorphous Ferromagnets. II. Long-Range Diffusion Measured by Degassing Experiments
(66.30; 61.40; 75.60; S1.1)
(a) Vol. **104**, No. 2, p. 619—626 (1987).

Hanmin Jin, Guanghua Guo, Ensheng Chen, Yong Zhu, Riuwen Zhao, Shinwen Li, and Fuming Yang
Magnetization Process in N–Fe–B Permanent Magnets near Remanence at Low Temperatures
(75.50; 75.60; S1.1; S1.4)
(a) Vol. **104**, No. 2, p. 799—804 (1987).

A. H. Wafik
*Effect of Cooling Rate on Barkhausen Jumps in Steel 110 F
(75.60; S1.1)
(a) Vol. **104**, No. 2, p. K133—K136 (1987).

75.70. Magnetic films and plates

H. PFEIFFER and R. HERGT
*Influence of Texture Formation on Properties of Films for Perpendicular Magnetic Recording
(68.55 and 75.70)
(a) Vol. **99**, No. 2, p. 549—558 (1987).

J. ČERMÁK, L. KALIVODA, and M. NEVŘIVA
*LPE Films of Magnetic Garnets with Bi Substitution
(68.55 and 75.70; S11.2)
(a) Vol. **100**, No. 1, p. 213—219 (1987).

W. ANDRÄ and H. DANAN
*Conditions for Perpendicular Magnetic Recording in Terms of the Wall-Displacement Model
(75.70; S.1.2)
(a) Vol. **100**, No. 1, p. 307—316 (1987).

T. BALCERZAK and J. P. JAKUBOVICS
*Studies of Fe–Ni and Fe–Ni–Gd Thin Films by Lorentz Microscopy
(75.70; 68.55; S1.1; S1.2)
(a) Vol. **101**, No. 1, p. 217—225 (1987).

U. HARTMANN
*Two-Dimensional Symmertrical Bloch-Type Boundaries in Thin Films
(75.70)
(a) Vol. **101**, No. 1, p. 227—232 (1987)..

E. M. ZUB and S. V. SUKHVALO
*Magnetic Properties of the Film System $Cu_xFe_{1-x}Cr_2S_4$
(75.70; S1.1)
(a) Vol. **101**, No. 1, p. K55—K60 (1987).

W. MACIESJEWSKI and A. DUDA
*Influence of the Surface Layers on the Curie Temperature of Compositionally Modulated Magnetic Films
(75.70)
(a) Vol. **101**, No. 1, p. K69—K72 (1987).

A. ZEGHIB, J. LE BAS, and C. VAUTIER
*Electrical and Magnetic Properties of Amorphous Ni_xAg_{1-x} Thin Films
(73.60 and 75.70; S1.2; S1.3)
(a) Vol. **101**, No. 2, p. 487—493 (1987).

D. I. SEMENTSOV and V. V. SIDORENKOV
*Ferromagnetic Resonance in Films with Angular Dispersion of Magnetic Anisotropy
(75.70; 76.50; S1.1; S1.2)
(a) Vol. **101**, No. 2, p. 555—562 (1987).

El Sayed Mehanna, S. Arajs, and H. F. Helbig
*Effects of Stresses on the Néel Temperature of Thin Polycrystalline Chromium Films
(73.60 and 75.70; S1.2)
(a) Vol. **101**, No. 2, p. K129—K133 (1987).

J. Baszyński
*Magnetic Surface Anisotropy of the Fe–Cu Interface in Superlattices
(75.50 and 75.70; 76.50; S1.1; S.1.2)
(b) Vol. **142**, No. 2, p. K165—K168 (1987).

R. Fabbri, G. Matteucci, and G. Pozzi
*Observations of Magnetic Domains in Thin Films by Electron Holography with a Commercial TEM with FEG
(61.16 and 75.70; S1.1)
(a) Vol. **102**, No. 2, p. K127—K129 (1987).

M. Jirsa, V. Kamberský, A. Kolodrubec, and M. Nevřiva
FMR Modes of the Bubble Domain Lattice in Garnet Films Magnetized along the Normal
(76.50; 75.60; 75.70; S11.2)
(a) Vol. **103**, No. 1, p. 219—230 (1987).

A. Duda
*The Influence of Next-Nearest-Neighbour Exchange Coupling on the Spin-Wave Resonance Spectrum in Uniaxial Ferromagnetic Thin Films
(75.70; 75.30; 76.50)
(b) Vol. **143**, No. 2, p. 709—716 (1987).

75.80. Magnetomechanical and magnetoelectric effects, magnetostriction

For galvanomagnetic effects, see 72.15 an 72.20
For magneto-optical effects, see 78.20

I. Retat
Plastic Deformation of Fe–Ni and $Fe_{65}(Ni_{1-x}Mn_x)_{35}$ between 30 and 460 K. Magnetically Induced Invar Hardening
(62.20; 75.80; S1.1; S1.2)
(a) Vol. **99**, No. 1, p. 121—130 (1987).

K. Terao and M. Shimizu
Electronic Structure and Magnetism of AFe_2 Compounds with C14 and C15 Laves Structures
(71.25 and 75.50; 75.80, S1.1)
(b) Vol. **139**, No. 2, p. 485—497 (1987).

O. Kohmoto and N. Yamaguchi
*Shift of Peak Position in μ–T Curves for Amorphous FeCo–SiB Alloys Having Zero Magnetostriction
(75.50 and 75.80; S1.1; S1.2)
(a) Vol. **100**, No. 1, p K61—K65 (1987).

A. Węgrzyn, L. T. Baczewski, and Z. Obuszko
*Magnetization and Magnetostriction of Iron-Rich $(Fe_{1-x}Co_x)_{90}Zr_{10}$ Metallic Glasses
(75.80; S1.1; S1.2)
(a) Vol. **100**, No. 2, p. K181—K184 (1987).

76. Magnetic resonances and relaxation in condensed matter; Mössbauer effect

76.20. General theory of resonances and relaxation

T. L. Buishvili and N. P. Fokina
*The Enhancement of the Low Frequency Susceptibility for Quadrupolar Splittings at Low Temperatures
(76.20)
(b) Vol. **140**, No. 1, p. K75—K78 (1987).

Yu. G. Semenov
Interaction of a Localized Spin Moment with Phonons Due to Exchange Scattering of Virtual Carriers in Semiconductors
(75.30; 75.80; 76.20; S8.1)
(b) Vol. **143**, No. 2, p. 717—726 (1987).

76.30. Electron paramagnetic resonance and relaxation

S. Waplak, J. Minge, and T. Krajewski
*EPR Studies of the Low-Temperature Phase Transition in $LiKSO_4$ by SO_4^- and CrO_4^{3-} Centres
(64.60 and 76.30; S11.1)
(a) Vol. **99**, No. 1, p. K27—K30 (1987).

G. S. Patrin, G. A. Petrakovskii, and V. V. Rudenko
*Photoinduced Change of Magnetic Resonance in $FeBO_3$ Single Crystals
(76.30; S11.1)
(a) Vol. **99**, No. 2, p. 619—623 (1987).

J. Kreissl and D. Backs
*EPR-Investigations on Mn-Doped CVD-Grown ZnS Thin Film Layers
(76.30; S8.11)
(a) Vol. **99**, No. 2, p. K117—K120 (1987).

D. Boulanger and R. Parrot
*Spin-Lattice Coupling Coefficients of Mn^{2+} in ZnSe
(76.30; S8.12)
(b) Vol. **140**, No. 1, p. K79—K82 (1987).

B. Ray, P. A. Monsef Mirzai, and J. W. Brightwell
Fluorescence and ESR in CaS: Mn^{2+}
(71.55; 76.30; 78.55; S8)
(a) Vol. **100**, No. 1, p. 233—237 (1987).

A. Hamaïda and J. Margerie
Measurement of Oscillator Strengths of "F_{2A}" Centres in $MeF_2:A^+$ (Me = Ca, Sr; A^+ = Li^+, Na^+, K^+)
(61.70; 76.30; 78.50; S9.12)
(b) Vol. **141**, No. 2, p. 391—398 (1987).

M. Onoda and H. Nagasawa
Phase Transition and Bipolaron Formation in Quasi-One-Dimensional Conductor β-$Ag_xV_2O_5$
(64.60 and 71.38; 76.30; 76.70; S11)
(b) Vol. **141**, No. 2, p. 507—516 (1987).

M. Herget, A. Hofstaetter, T. Nickel, and A. Scharmann
Reorientation of Hole Centers in $CaWO_4$ Monitored by EPR and Optical Absorption
(71.38 and 71.70; 76.30; 78.50; S11.1)
(b) Vol. **141**, No.2, p. 523—528 (1987).

C. Weber, M. Wagner, and E. Sigmund
*Flexibility Options in Valency-Mixing Jahn-Teller Systems
(71.70 and 76.30; S1.4)
(b) Vol **141**, No. 2, p. 529—539 (1987).

Zheng Wen-Chen
*Separation of Cubic and Axial Parts of the Crystal Field from Hydrostatic and Uniaxial Pressure Dependences of the Parameter a_0
(76.30; 71.70; S11.1)
(b) Vol. **141**, No. 2, p. K121—K124 (1987).

T. N. Sairam and S. V. Bhat
*EPR Study of Mn^{2+} Doped Dicalcium Barium Propionate Single Crystals under Hydrostatic Pressure
(76.30; 64.70; S12)
(b) Vol. **141**, No. 2, p. K133—K138 (1987).

M. Das and A. K. Pal
EPR Studies on Phase Transitions in Copper Tetramethylammonium Bromide
(64.60; 76.30; S9)
(a) Vol. **101**, No. 2, p. 403—408 (1987).

J. Minge, S. Waplak, and T. Krajewski
EPR Study of Low-Temperature Phase Transitions in NH_3^+-Doped $LiKSO_4$
(64.70; 76.30; S11.1)
(a) Vol. **101**, No. 2, p. 427—435 (1987).

Zhou Yi-Yang
EPR Study of Cu^{2+} Ion in $K_2Cd(SO_4)_2 \cdot 6\,H_2O$
(71.70; 76.30; S11)
(b) Vol. **142**, No. 1, p. 229—235 (1987).

P. Wünsche, R. Kunze, and M. Wäsche
Activation Energy Distribution in Poly-(Methylmethacrylate–Methacrylic Acid) Estimated by EPR and Thermoluminescence
(65.90; 76.30; 78.60; S12)
(a) Vol. **103**, No. 1, p. 145—152 (1987).

E. K. Sadykov and A. I. Skvortsov
*On the Theory of Double Gamma Electronic Magnetic Resonance in Paramagnetic Systems
(76.30; 76.80)
(b) Vol. **143**, No. 2, p. 699—707 (1987).

A. Stesmans
*Interaction between Localized and Delocalized States Indicated from Temperature-Dependent ESR Study of Submetallic P-Doped Si
(76.30; 71.25; S5.11)
(b) Vol. **143**, No. 2, p. 733—740 (1987).

Y. Ravi Sekhar and H. Bill
*^{2}S Ground State Ions in Ionic Crystals: ESR Results
(76.30; 61.70; S9.12; S10)
(b) Vol. **143**, No. 2, p. K111—K114 (1987).

O. Thiemann and O. F. Schirmer
*Microwave Electric Field Induced ESR of Ni^{2+} in $LiNbO_3$
(76.30; S11.1)
(b) Vol. **143**, No. 2, p. K115—K118 (1987).

B. Milsch and F. Kerbe
*EPR of Chromium and Manganese Ions in Beryllium-Oxide Ceramics
(76.30; S10)
(a) Vol. **103**, No. 2, p. K141—K144 (1987).

Xiao-Peng He and Mao-Lu Du
*An EPR Study of the Structure of the Cr^{3+}–Vacancy Centre in MgO
(76.30; S10.1)
(b) Vol. **144**, No. 1, p. K51—K56 (1987).

Yu. N. Orlov, V. I. Aleshin, G. E. Arkhangelskii, V. E. Bozhevolnov, Yu. V. Voronov, L. N. Ivanov, V. V. Karelin, and E. I. Saunin
Concentration Peculiarities of a Set of Impurity Centres in Fluorite Crystals Activated by Gadolinium
(61.70; 76.30; 78.60; S9.15)
(b) Vol. **144**, No. 2, p. K77—K80 (1987).

N. Guskos
*An Interpretation of the Electric Field Effect on Cubic Gd^{3+} Centers in BaF_2 Single Crystals
(76.30; 71.70; S9.12)
(b) Vol. **144**, No. 2, p. K139—K142 (1987).

76.40. Diamagnetic and cyclotron resonances

76.50. Ferromagnetic, antiferromagnetic, and ferrimagnetic resonances; spin-wave resonance

(see also 75.30. Spin waves)

D. I. SEMENTSOV and V. V. SIDORENKOV
Ferromagnetic Resonance in Films with Angular Dispersion of Magnetic Anisotropy
(75.70; 76.50; S1.1; S1.2)
(a) Vol. **101**, No. 2, p. 555—562 (1987).

J. BASZYŃSKI
Magnetic Surface Anisotropy of the Fe–Cu Interface in Superlattices
(75.50 and 75.70; 76.50; S1.1; S1.2)
(b) Vol. **142**, No. 2, p. K165—K168 (1987).

M. JIRSA, V. KAMBERSKÝ, A. KOLODRUBEC, and M. NEVŘIVA
*FMR Modes of the Bubble Domain Lattice in Garnet Films Magnetized along the Normal
(76.50; 75.60; 75.70; S11.2)
(a) Vol. **103**, No. 1, p. 219—230 (1987).

A. DUDA
The Influence of Next-Nearest-Neighbour Exchange Coupling on the Spin-Wave Resonance Spectrum in Uniaxial Ferromagnetic Thin Films
(75.70; 75.30; 76.50)
(b) Vol. **143**, No. 2, p. 709—716 (1987).

H. BOSSE and H. GÄRTNER
*A Systematic Ferromagnetic Resonance Study of Fe-, Ni-, and Cr-Ion Plated Amorphous Ribbons
(75.50 and 76.50; S1.1; S1.2)
(a) Vol. **103**, No. 2, p. 565—570 (1987).

76.60. Nuclear magnetic resonance and relaxation

H. FIGIEL, A. C. BARATA, and A. P. GUIMARÃES
*^{59}Co NMR and Nuclear Magnetic Relaxation Study of the Magnetic Superconductor Y_9Co_7
(76.60; S1)
(b) Vol. **139**, No. 1, p. 311—314 (1987).

F. ENGELKE, D. MICHEL, W. WINDSCH, and J. PETERSSON
*^{2}D NMR Study near the Ferroelectric Phase Transition of TSCC
(76.60 and 77.80; S12)
(b) Vol. **139**, No. 2, p. 609—617 (1987).

B. M. BOLOTOV, D. A. AKSENOV, A. I. POLYAKOV, and S. P. SENSHIN
*NMR Line Shape Function Analysis in Solids
(76.60; S4; S9.11; S9.12)
(b) Vol. **139**, No. 2, p. K137—K140 (1987).

R. GOC and D. FIAT
*NMR Powder Spectrum Simulation for Nuclei with $I > 1/2$
(76.60)
(b) Vol. **140**, No. 1, p. 243—250 (1987).

76.70. Magnetic double resonances and cross effects

A. Davituliani, M. Kiknadze, R. Mirianshvili, and T. Sanadze
*RFDS Studies of Trigonal Er^{3+} Centres in BaF_2 Single Crystals
(76.70; S9.12)
(b) Vol. **140**, No. 1, p. K69—K73 (1987).

M. Onoda and H. Nagasawa
Phase Transition and Bipolaron Formation in Quasi-One-Dimensional Conductor β-$Ag_xV_2O_5$
(64.60 and 71.38; 76.30; 76.70; S11)
(b) Vol. **141**, No. 2, p. 507—516 (1987).

V. P. Meiklyar, A. E. Usachev, and Yu. V. Yablokov
*ENDOR of the Yb^{3+} Ion in Cs_2NaYCl_6 Crystal
(76.70; S11)
(b) Vol. **141**, No. 2, p. K129—K132 (1987).

76.80. Mössbauer effect; other γ-ray spectroscopy

A. Ostrasz and M. Szuskiwicz
3d-Like Electron Localization and Charge Transfer in Hydrogenated V–Fe Alloys
(71.25; 61.55; 76.80; S1.1; S1.2)
(b) Vol. **139**, No. 1, p. 205—211 (1987).

J. Li and B. X. Liu
*CEMS to Characterize the Ion Induced Fe–Mo Metastable Phase
(64.60 and 76.80; S1; S1.1)
(a) Vol. **99**, No. 1, p. K31—K34 (1987).

J. Fontcuberta
Lattice Dynamics Study of Polycrystalline ^{57}CoO by Mössbauer Spectroscopy
(63.20; 76.80; S10.1)
(b) Vol. **139**, No. 2, p. 379—386 (1987).

A. R. Mkrtchyan, R. G. Gabrielyan, A. H. Martirossian, and A. Sh. Grigoryan
*γ-Quantum Resonance Scattering in a Plane-Parallel Isotropic Layer
(76.80)
(b) Vol. **139**, No. 2, p. 583—595 (1987).

E. I. Vapirev, K. I. Bourin, I. D. Christoskov, L. T. Tsankov, and P. S. Kamenov
*Random Motion Studied by Time Delayed Mössbauer Spectroscopy
(76.80)
(b) Vol. **142**, No. 1, p. 279—285 (1987).

E. Kolawa, T. Tyliszczak, B. D. Sawicka, J. A. Sawicki, and M. Drwiega
Studies of Local Structure and Thermal Annealing Processes in Implanted Iron–Gadolinium Alloys by Conversion Electron Mössbauer Spectroscopy
(61.40; 61.70; 76.80; S1.1; S1.4)
(a) Vol. **102**, No. 1, p. 181—192 (1987).

A. Ostrasz and M. Szuskiewicz
*Comment on the Magnetic Moments Formation in Hydrogenated V–Fe Alloys
(76.80; S1.2)
(b) Vol. **142**, No. 2, p. K173—K176 (1987).

J. K. Srivastava, S. Muralidhara Rao, R. Nagarajan, and R. Vijayaraghavan
*Mössbauer Effect, Ionic Conductivity, and Ionic Thermoconductivity Studies of Ionic Spin Fluctuations in the Magnetically Ordered $Fe^{3+}:NiTiO_3$ System
(76.80; 66.30; S10.15)
(a) Vol. **102**, No. 2, p. 759—768 (1987).

H. Jex and U. Müller
X-Ray Brillouin Scattering of Mössbauer Radiation by an Oscillating Quartz Crystal
(63.20 and 78.35; 76.80; S10.1)
(b) Vol. **143**, No. 1, p. 317—324 (1987).

I. Śledzińska
*Electric Field Gradient in Ferrous Oxalate Dihydrate
(76.80; S11)
(b) Vol. **143**, No. 1, p. K63—K66 (1987).

A. Hauet, J. Teillet, B. Hannoyer, and M. Lenglet
*Mössbauer Study of Co and Ni Ferrichromites
(76.80; 75.50; S11.2)
(a) Vol. **103**, No. 1, p. 257—261 (1987).

E. K. Sadykov and A. I. Skvortsov
On the Theory of Double Gamma Electronic Magnetic Resonance in Paramagnetic Systems
(76.30; 76.80)
(b) Vol. **143**, No. 2, p. 699—707 (1987).

Cs. Szeles, K. Süvegh, Z. Homonnay, and A. Vértes
Positron Lifetime and Mössbauer Spectroscopy Study of Vacancy–Tin Interaction in Dilute Al–Sn Alloys
(61.70; 76.80; 78.70; S4)
(a) Vol. **103**, No. 2, p. 397—401 (1987).

A. I. C. Persiano and R. D. Rawlings
*A Mössbauer Investigation of Equiatomic FeCo with Vanadium and Niobium Additions
(75.50 and 76.80; S1.1; S1.2)
(a) Vol. **103**, No. 2, p. 547—556 (1987).

T. STAPIŃSKI, E. JAPA, and J. ZUKROWSKI
*^{119}Sn Mössbauer Investigation of Cadmium–Tin–Oxide Thin Films
(68.55 and 76.80; S10.1)
(a) Vol. **103**, No. 2, p. K93—K98 (1987).

C. GIOVANELLA, M. IANNUZZI, F. GAUZZI, G. PETROCCO, G. PRINCIPI, and P. Q. ZHANG
*Production of Structurally and Magnetically Disordered $Fe_{20}Cu_{80}$ Metastable Alloys
(76.80; S1.1; S1.2)
(a) Vol. **103**, No. 2, p. K125—K129 (1987).

G. VOGL
On the Motion of Solute Atoms in Cages and Loops in Metal Lattices Studied with Nuclear Methods
(61.70; 66.30; 76.80; S1; S1.3; S4)
(b) Vol. **144**, No. 1, p. 259—269 (1987).

F. AUBERTIN, G. L. WHITTLE, S. J. CAMPBELL, and U. GONSER
Hydrogen-Induced Magnetism in Zr_3Fe Hydrides
(75.30; 61.10; 76.80; S1.1)
(a) Vol. **104**, No. 1, p. 397—402 (1987)

I. DÉZSI, H. ENGELMANN, U. GONSER, and G. LANGOUCHE
Cobalt–Germanium Structures Investigated by Mössbauer Spectroscopy
(61.55; 61.80; 76.80; S1.16; S5.12)
(a) Vol. **104**, No. 1, p. 417—426 (1987).

M. MIGLIERINI and J. SITEK
*Low Temperatures Mössbauer Study of Amorphous $Fe_{83}B_{17}$
(76.80; 75.25; S1.1)
(a) Vol. **104**, No. 2, p. K137—K139 (1987).

76.90. Other topics in magnetic resonances and relaxation

M. SOUIRI, A. GOLTZENE, and C. SCHWAB
The Environment of Muonium in Solids. An Empirical Approach
(76.30; 76.90)
(b) Vol. **142**, No. 1, p. 271—278 (1987).

S. N. KONYAEV, A. I. SHUSHIN, L. I. KOLESNIKOVA, M. M. TRIBEL, and E. L. FRANKEVICH
Fluorescence-Detected Low Field Resonances in T–T Annihilation in the A-TCNB Single Crystal
(71.35; 76.90; 78.55; S12)
(b) Vol. **142**, No. 2, p. 461—471 (1987).

H. PINKVOS and CH. SCHWINK
Muon Spin Rotation Spectroscopy in Metallic Spin Glass Systems
(75.40; 76.90)
(b) Vol. **144**, No. 1, p. 351—360 (1987).

77. Dielectric properties and materials
(for conductivity phenomena, see 72.20 and 72.80)

77.20. Permittivity

L. Sirdeshmukh, G. Sathaiah, and P. Devi
*Dielectric Properties of the KCl–KBr Mixed Crystal System
(77.20 and 77.40; S9.15)
(a) Vol. **99**, No. 2, p. 631—639 (1987).

M. K. Teng, J. F. Vittori, M. Massot, and S. Ziolkiewicz
*Dielectric and Ferroelectric Properties in $As_xSb_{1-x}SI$ Mixed Crystals
(77.20 and 77.80; S9.16)
(a) Vol. **99**, No. 2, p. 641—647 (1987).

G. M. S. Srivastava, R. K. Sinha, and J. Shanker
*Interrelationship between Higher-Order Elastic Constants and Pressure Derivatives of Dielectric Constants for Ionic Crystals
(62.20 and 77.20; S9.11)
(b) Vol. **140**, No. 1, p. 89—95 (1987).

J. Monecke
*Material Constants of Composites. I. General Theory for the Dielectric Constant
(77.20)
(b) Vol. **140**, No. 1, p. 279—287 (1987).

A. Batana and J. Faour
*Study of the Pressure Dependence of Dielectric Properties of Ionic Crystals with the Exchange-Charge Model
(77.20; 9.11)
(b) Vol. **140**, No. 1, p. K87—K90 (1987).

A. K. Pradhan and R. N. P. Choudhary
*Dielectric Properties of $SmAsO_4$ and $YAsO_4$
(77.20 and 77.40; S11.1)
(a) Vol. **100**, No. 1, p. 323—326 (1987)

H. Kresse and P. Rabenstein
*Geometrical Aspects and Dipole–Dipole Interaction in a Liquid Crystalline Mixture
(61.25 and 77.20; S12)
(a) Vol. **100**, No. 1, p. K83—K85 (1985).

H. Kresse, W. Weissflog, C. Nagabhushan, and S. Shashidhar
*Dipole–Dipole Interactions in Swallow-Tailed Liquid Crystalline Mixtures
(77.20; S12)
(a) Vol. **101**, No. 1, p. K77—K79 (1987).

A. Vohra, R. Verma, K. N. Lakshminarayan, and K. K. Srivastava
*Dielectric Parameters for Planar Samples of Non-Conducting and Semiconducting Solid Powders at Microwave Frequencies
(77.20; S8)
(a) Vol. **101**, No. 1, p. K81—K85 (1987).

77.30. Polarization and depolarization effects

L. Wendler and R. Pechstedt
Dynamical Screening, Collective Excitations, and Electron-Phonon Interaction in Heterostructures and Semiconductor Quantum Wells. Application to Double Heterostructures
(71.38 and 71.45; 68.30; 73.40; 77.30; 78.65; S7.12; S7.15)
(b) Vol. **141**, No. 1, p. 129—150 (1987).

J. Prakash and A. K. Nishad
*Prediction of ITC Parameters at the Required Heating Rate
(77.30 and 77.40)
(a) Vol. **101**, No. 2, p. 629—637 (1987).

M. L. Mukherjee and R. Capelletti
X-Ray Induced Relaxation Band Due to Higher Aggregates in KCl:Pb
(61.70; 77.30; 78.50; S9.11)
(b) Vol. **142**, No. 2, p. 361—366 (1987).

L. N. Kantorovich
Multipole Theory of the Polarization of Solids by Point Defects. IV. Electronic and Spatial Structure of Single Electronic and Hole Centers and Their Pairs in LiF, KCl Crystals
(71.55; 77.30; S9.11)
(b) Vol. **144**, No. 2, p. 719—726 (1987).

77.40. Dielectric loss and relaxation

L. Sirdeshmukh, G. Sathaiah, and P. Devi
*Dielectric Properties of the KCl-KBr Mixed Crystal System
(77.20 and 77.40; S9.15)
(a) Vol. **99**, No. 2, p. 631—639 (1987).

T. V. Narasaiah and R. N. P. Choudhary
Dielectric and Thermal Properties of TlD_2PO_4
(77.80; 77.40; S11.1)
(a) Vol. **100**, No. 1, p. 317—321 (1987).

A. K. Pradhan and R. N. P. Choudhary
*Dielectric Properties of $SmAsO_4$ and $YAsO_4$
(77.20 and 77.40; S11.1)
(a) Vol. **100**, No. 1, p. 323—326 (1987).

U. Subramanian and M. L. Mukherjee
*Dielectric Studies of CaF_2:Nd Single Crystals from RT to 350 °C
(77.20 and 77.40; S9.12)
(a) Vol. **101**, No. 2, p. 577—581 (1987).

J. PRAKASH and A. K. NISHAD
*Prediction of ITC Parameters at the Required Heating Rate
(77.30 and 77.40)
(a) Vol. **101**, No. 2, p. 629—637 (1987).

K. BHOI, K. V. RAO, and A. V. KRISHNA RAO
*Study of Space Charge Processes in High Electric Field Treated Rutile Single Crystals Using Dielectric Measurements
(77.20 and 77.40; S10.1)
(a) Vol. **102**, No. 2, p. K131—K136 (1987).

M. R. DE LA FUENTE, M. A. PÉREZ JUBINDO, J. D. SOLIER, and M. J. TELLO
*The Memory Effect in Empirical Descriptions of Dielectric Relaxation
(77.40)
(b) Vol. **143**, No. 1, p. 299—306 (1987).

Y. R. REDDY and L. SIRDESHMUKH
*Dielectric Behaviour of $Bi_{12}SiO_{20}$
(77.20 and 77.40; S10)
(a) Vol. **103**, No. 2, p. K157—K160 (1987).

B. S. BERRY and W. C. PRITCHET
*Computer Modeling of Relaxation Peaks Involving a Gaussian Distribution of Activation Energies
(62.40 and 77.40)
(b) Vol. **144**, No. 1, p. 375—384 (1987).

M. R. DE LA FUENTE, M. A. PÉREZ JUBINDO, R. BALDA, and M. J. TELLO
*Influence of Dipolar Interactions on the Dynamical Dielectric Response
(77.40; 77.80)
(b) Vol. **144**, No. 2, p. 825—833 (1987).

B. NARASIMHA, K. V. RAO, and R. N. P. CHOUDHARY
*Study on Dielectric Dispersion of Neodymium Phosphate Ceramics
(77.20 and 77.40; S11.1)
(a) Vol. **104**, No. 2, p. 849—853 (1987).

77.50. Dielectric breakdown and space-charge effects

V. P. GRÍBKOWSKII, A. L. GURSKÍI, G. A. PASHKEVICH, and G. P. YABLONSKII
*Crystallographic Orientation of Electric Discharges in TeO_2 and $LiNbO_3$ Monocrystals
(77.50; 63.20; S10.1; S11.1)
(a) Vol. **103**, No. 2, p. K153—K156 (1987).

77.55. Dielectric thin films

E. G. SALMAN, G. V. PEROV, and V. N. VERTOPRAKHOV
Ionic Current Study in Thermally Grown Silicon Dioxide Films on Polycrystalline Silicon
(73.40 and 77.30; 77.55; S10.1)
(a) Vol. **99**, No. 2, p. 537—542 (1987).

R. SINGH, J. M. KELLER, and S. C. DATT
*Thermally Stimulatd Current and Charge Decay Studies in Polypropylene
(77.55; S12)
(a) Vol. **100**, No. 1, p. 143—147 (1987).

PAN JIN-SHENG
*Electron–Phonon Interaction in Dielectric Interface Systems
(77.55; 68.30; 71.38)
(b) Vol. **140**, No. 2, p. 421—436 (1987).

M. V. R. REDDY, J. S. KUMAR, K. N. REDDY, and U. V. S. RAO
*Dielectric Properties of $NaYF_4$ Thin Films
(77.55; S9.16)
(a) Vol. **102**, No. 1, p. 321—326 (1987).

N. SASI, C. BALASUBRAMANIAN, and S. K. NARAYANDASS
Structure, Dielectric, and AC Conduction Studies on Germanium Dioxide Thin Films
(68.55 and 73.60; 77.55; S10.1)
(a) Vol. **103**, No. 2, p. 475—480 (1987).

77.70. Pyroelectric and electrocaloric effects

Y. A. BADR, S. B. EL-GIZIRI, and T. M. HAMMAD
On the Effect of Doping on the Mechanism of Phase Transition of Some Ionic Inorganic Compounds
(64.70 and 77.80; 72.80; 77.70; S11.1)
(a) Vol. **100**, No. 2, p. 467—472 (1987).

T. ASAJI, M. TAYA, and D. NAKAMURA
*Pyroelectricity of Meta-Substituted Nitrobenzenes m-$XC_6H_4(NO_2)$ ($X = NH_2$, Br, Cl)
(77.70; S12)
(a) Vol. **102**, No. 2, p. 815—818 (1987).

B. WYNCKE and F. BREHAT
Ferroelectric Properties of a KNO_3 Single Crystal. Application to the Detection of Thermal Radiation
(77.80; 77.70; S11.1)
(a) Vol. **104**, No. 2, p. 873—877 (1987).

77.80. Ferroelectricity and antiferroelectricity

B. V. Egorov, I. B. Egorova, and M. A. Krivoglaz
*The Electronic Heterogeneous States in Ferroelectric Semiconductors
(71.25 and 77.80; 65.50; 77.30)
(b) Vol. **139**, No. 1, p. 173—184 (1987).

J. M. Wesselinowa
*Deuteration Dependence of the Central Peak in KDP-Type Ferroelectrics
(77.80)
(b) Vol. **139**, No. 1, p. 329—335 (1987).

S. Ahmed, A. M. Shamah, R. Kamel, and Y. Badr
Structural Changes of $(NH_4)_2SO_4$ Crystals
(64.70; 77.80; S11.1)
(a) Vol. **99**, No. 1, p. 131—140 (1987).

F. Engelke, D. Michel, W. Windsch, and J. Petersson
*^{2}D NMR Study near the Ferroelectric Phase Transition of TSCC
(76.60 and 77.80; S12)
(b) Vol. **139**, No. 2, p. 609—617 (1987).

M. K. Teng, J. F. Vittori, M. Massot, and S. Ziolkiewicz
*Dielectric and Ferroelectric Properties in $As_xSb_{1-x}SI$ Mixed Crystals
(77.20 and 77.80; S9.16)
(a) Vol. **99**, No. 2, p. 641—647 (1987).

T. V. Narasaiah and R. N. P. Choudhary
*Dielectric and Thermal Properties of TlD_2PO_4
(77.80; 77.40; S11.1)
(a) Vol. **100**, No. 1, p. 317—321 (1987).

K. Biedrzycki and B. Kosturek
*Thermally Stimulated Electron Emission from Triglycine Sulphate Crystals with Well-Defined Initial Domain Structure
(77.80 and 79.40; S12)
(a) Vol. **100**, No. 1, p. 327—330 (1987).

J. K. Rath and S. Radhakrishna
*EPR Studies of MoO_4^{3-} Centres in KDP Single Crystals
(77.80; 76.30; S11.1)
(a) Vol. **100**, No. 1, p. 331—336 (1987).

G. Koschek and E. Kubalek
*Rasterelektronenmikroskopische Direktabbildung von ferroelektrischen Domänen in $BaTiO_3$-Keramiken durch Sekundärelektronen und Kathodolumineszenz
(Scanning Electron Microscope Direct Imaging of Ferroelectric Domains in $BaTiO_3$ Ceramics by Secondary Electrons and Cathodoluminescence)
(77.80; 78.60)
(a) Vol. **100**, No. 1, p. 355—368 (1987).

Y. A. Badr, S. B. El-Giziri, and T. M. Hammad
*On the Effect of Doping on the Mechanism of Phase Transition of Some Ionic Inorganic Compounds
(64.70 and 77.80; 72.80; 77.70; S11.1)
(a) Vol. **100**, No. 2, p. 467—472 (1987).

W. Osak and K. Tkacz
*Long-Lasting Relaxation Currents in TGS
(77.80; S12)
(a) Vol. **100**, No. 2, p. 667—672 (1987).

H.-E. Müller, G. Völkel, W. Brunner, P. Cevc, and I. N. Kurkin
*Electron Spin Relaxation in Ferroelastic $KH_3(SeO_3)_2$
(76.30 and 77.80; S11.1)
(b) Vol. **141**, No. 1, p. 343—349 (1987).

V. I. Voronkova and V. K. Yanovskii
The Structure and Ferroelectric Properties of Layered Compounds in the System $Bi_4Ti_3O_{12}$–Bi_2WO_6
(61.60; 64.70; 77.80; S10.15)
(a) Vol. **101**, No. 1, p. 45—50 (1987).

W. Osak and K. Tkacz
*An Improved Way of Obtaining *I–U* Characteristics of Ferroelectrics
(77.80; S12)
(a) Vol. **101**, No. 1, p. 263—266 (1987).

J. M. Wesselinowa
*Relaxation Time for KDP- and DKDP-Type Ferroelectrics
(77.80; S11.1)
(b) Vol. **142**, No. 1, p. K85—K89 (1987).

R. Styrowiec and W. A. Sysło
*The Simultaneous Effect of Electric Field and Mechanical Stress on the Domain Walls of Rochelle Salt Crystals
(77.80; S11)
(a) Vol. **102**, No. 1, p. K49—K51 (1987).

S. S. Bawa, A. M. Biradar, and S. Chandra
*Molecular Reorientation Process and Non-Linear Polarization Reversal Current in Helicoidal Ferroelectric Crystal DOBAMBC
(77.80; S12)
(a) Vol. **102**, No. 2, p. 829—834 (1987).

M. A. Gaffar, M. M. Mebed, and A. Abu El-Fadl
*The Specific Heat of Pure and Doped Triglycine Sulphate Single Crystals
(65.40 and 77.80; S12)
(a) Vol. **103**, No. 2, p. 459—466 (1987).

78. Optical properties and condensed-matter spectroscopy and other interactions of matter with particles and radiation

(for phonon spectra, see 63)

78.20. Optical properties and materials

J. BULLOT and M. P. SCHMIDT
Physics of Amorphous Silicon–Carbon **(Review Article)**
(61.40; 71.20; 71.55; 72.40; 73.60; 78.20; 78.55; 78.60; S6)
(b) Vol. **143**, No. 2, p. 345—418 (1987).

A. MASŁOWSKA
The Burstein Absorption Edge Structure in Lead Chalcogenides
(71.25; 71.38; 78.20; S8)
(b) Vol. **139**, No. 1, p. 161—171 (1987).

S. F. CHEHAB and J. C. WOLLEY
Energy Gap and Valence Band Structure of $Cd_{1-x}Mn_xTe_{1-y}Se_y$ Alloys
(71.25; 78.20; S8.15)
(b) Vol. **139**, No. 1, p. 213—222 (1987).

G. MANZKE, K. HENNEBERGER, and V. MAY
Many-Exciton Theory for Multiple Quantum-Well Structures
(71.35; 78.20; S7.12; S7.15)
(b) Vol. **139**, No. 1, p. 233—239 (1987).

R. SOMMERFELDT, R. A. RUPP, H. VORMANN, and E. KRÄTZIG
Thermal Fixing of Volume Phase Holograms in $LiNbO_3$:Cu
(66.30; 78.20; S11.1)
(a) Vol. **99**, No. 1, p. K15—K18 (1987).

T. KESSLER, R. MARKUS, H. NAHME, and N. SCHWENTNER
*Nonlinear Excitation of Pure and Doped Rare Gas Crystals
(78.20; 71.35; 78.55; S5.2)
(b) Vol. **139**, No. 2, p. 619—625 (1987).

YU. M. SEIDOV and A. M. SULEIMANOV
To the Theory of Two-Magnon Absorption of Light by Surface Collective Excitations in the Singlet–Triplet Ferromagnets
(71.70; 75.30; 78.20; S1.4)
(b) Vol. **140**, No. 1, p. 149—154 (1987).

G. CZAJKOWSKI and P. SCHILLAK
*Interference Spectra of Semiconducting Slabs at Excitonic Resonances Studied by Stahl's Coherent Wave Approach to Band Gap Electrodynamics
(78.20; 71.35; S8.11; S8.12)
(b) Vol. **140**, No. 1, p. 273—278 (1987).

M. Załużny and W. Okulski
*On the Interband Absorption in Size-Quantized Bi Films
(78.20 and 78.65; 71.25; S4)
(b) Vol. **141**, No. 2, p. 427—434 (1987).

Cz. Koepke
*Cavityless Optical Bistability in Thin Sprayed CdS Films
(78.20; S8.11)
(b) Vol. **141**, No. 2, p. K139—K142 (1987).

M. I. Strashnikova, V. Ya. Reznichenko, and V. V. Cherny
*Size Effect in the Refractive Index Dispersion of Thin CdS Crystals
(78.20; S8.11)
(b) Vol. **141**, No. 2, p. K153—K156 (1987).

R. I. Joseph and M. E. Thomas
*Differential Reflection Spectroscopy: A Direct Determination of Long Wavelength Optical Mode Frequencies in Polar Crystals
(78.20; S10.1)
(b) Vol. **141**, No. 2, p. K163—K166 (1987).

S. N. Toubektsis, J. A. Kalomiros, L. V. Giobliakis, K. M. Paraskevopoulos, A. N. Anagnostopoulos, and E. K. Polychroniadis
On the Growth and Characterization of the System $TlBiSe_2$–$TlBiTe_2$
(61.60; 72.80; 78.20; S8.16)
(a) Vol. **101**, No. 2, p. 355—360 (1987).

J. Berdowski
*The Influence of Isotope Effects on the Light Diffraction by Surface Acoustic Waves at Perpendicular Interaction in DADA Crystals
(78.20; S11)
(a) Vol. **102**, No. 1, p. 399—409 (1987).

G. Jungk
*On the Emissivity of Composite Materials
(78.20)
(a) Vol. **102**, No. 1, p. 411—416 (1987).

P. Lukáš and J. Kulda
*On the Reflectivity of Deformed Absorbing Crystals
(78.20)
(a) Vol. **102**, No. 1, p. K57—K60 (1987).

K. Takarabe, H. Kawamura, and K. Wakamura
*Electronic Band Structure and Optical Anisotropy in InS. I. Experimental Study
(71.25 and 78.20; S8)
(b) Vol. **142**, No. 2, p. 605—615 (1987).

F. Lukeš
*Temperature Dependence of Ellipsometric Angles of Silicon
(78.20; 68.45; S5.11)
(a) Vol. **102**, No. 2, p. 803—814 (1987).

R. Bukowski and Z. Kleszczewski
*Non-Stationary Stimulated Acousto-Optical Interactions in Solid Dielectrics
(78.20)
(b) Vol. **143**, No. 1, p. 281—287 (1987).

J. E. Lewis
*Optical Properties of the Alloy System SnTe–GeTe from Reflectance Measurements
(78.20; 71.25; 78.30; S8)
(b) Vol. **143**, No. 1, p. 307—315 (1987).

J. M. Dance, J. P. Chaminade, J. M. Moutou, and M. Pouchard
*Spectroscopic and Magnetic Characterization of the Copper–Tungsten Oxyfluoride $CuWO_3F_2$.
(75.30 and 76.30; 78.20; S9.16)
(b) Vol. **143**, No. 1, p. K57—K61 (1987).

M. Kretzschmar, F. Henneberger, H. Rossmann, and I. Haddad
*Dynamical Effects in Increasing Absorption Optical Bistability of Thermal Origin
(78.20; S8.11)
(b) Vol. **143**, No. 1, p. K71—K76 (1987).

B. Esser and A. Selloni
*Multiphonon Coupling of Optical Transitions in Disordered Systems and Inverse Participation Ratio
(78.20)
(b) Vol. **143**, No. 1, p. K77—K81 (1987).

H. Röppischer, R. Goldhahn, J. A. Bumai, and D. S. Domanevskii
*Electroreflectance Depth Profiling and Hall Measurements of Thermally Converted GaAs Layers
(73.60 and 78.20; S7.12)
(a) Vol. **103**, No. 1, p. 161—166 (1987).

P. Górski and W. Kucharczyk
*The Quadratic Electrooptic Effect in KDP and ADP Crystals
(78.20; S11.1)
(a) Vol. **103**, No. 1, p. K65—K67 (1987).

K. Takarabe, H. Kawamura, and K. Wakamura
*Electronic Band Structure and Optical Anisotropy in InS. II. Theoretical Study
(71.25 and 78.20; S8)
(b) Vol. **143**, No. 2, p. 519—538 (1987).

C. Medrano P., P. Günter, and H. Arend
*Noncentrosymmetry Observed in $CaC_4H_4O_6 \cdot 4\,H_2O$ Crystals by Nonlinear Optical Measurements
(78.20; S12)
(b) Vol. **143**, No. 2, p. 749—754 (1987).

H. Puff and H. W. Streitwolf
*Optical Absorption of the Soliton Lattice
(78.20; 71.45; S12)
(b) Vol. **143**, No. 2, p. 765—775 (1987).

J. L. SUBTIL, P. LAPORTE, R. REININGER, and V. SAILE
*VUV Optical Constants of Liquid and Solid Xenon at the Triple Point
(78.20; 71.35; S5.2)
(b) Vol. **143**, No. 2, p. 783—793 (1987).

H. BOUDRIOT, K. DEUS, R. GRÜNDLER, H. A. SCHNEIDER, and T. STÖCKERT
*Optical Properties of $ZnSiAs_2$ Single Crystalline Thin Films from Electron Energy Loss Measurements
(78.20 and 79.20; 71.25; S7.16)
(b) Vol. **143**, No. 2, p. K125—K130 (1987).

O. FOLTIN
*Oblique Incidence Specular Reflectivity of Isotropic Semiconductors near Three-Dimensional Critical Points
(78.20; 71.20)
(b) Vol. **143**, No. 2, p. K151—K154 (1987).

F. Y. M. AL-EITHAN, M. C. AL-EDANI, and A. K. ABASS
*Optical Band Gap of Triglycine Selenate (TGSe) Single Crystals
(78.20; S12)
(a) Vol. **103**, No. 2, p. 571—574 (1987).

T. WILSON and P. D. PESTER
*Information Depth in Optical Beam Induced Examination of Semiconductor Materials and Devices
(78.20; 78.55)
(a) Vol. **103**, No. 2, p. 631—636 (1987).

G. A. MEDVEDKIN, YU. V. RUD, and M. A. TAIROV
*Optical Band Edge Absorption in $CuInSe_2$
(78.20; S8.16)
(b) Vol. **144**, No. 2, p. 809—816 (1987).

M. JOUANNE, C. JULIEN, and M. BALKANSKI
*Polarization Studies of Polar Phonons of InSe
(63.20 and 78.20; S8)
(b) Vol. **144**, No. 2, p. K147—K150 (1987).

S. A. STEPANOV, I. S. EDELMAN, T. A. KIM, G. T. PETROVSKII, and G. V. POPOV
*Properties of Magnetically Ordered Microparticles in Borate Glasses
(75.50 and 78.20; 61.40; S10.15)
(a) Vol. **104**, No. 2, p. 805—814 (1987).

P. HERTEL, K. H. RINGHOFER, and R. SOMMERFELDT
*Theory of Thermal Hologram Fixing and Application to $LiNbO_3$:Cu
(72.40 and 78.20; S11.1)
(a) Vol. **104**, No. 2, p. 855—862 (1987).

F. KOCH and O. OELGART
Reabsorbed Recombination Radiation and Electron Beam Induced Barrier Current in n-Type GaAs
(73.40; 78.20; 78.55; S7.12)
(a) Vol. **104**, No. 2, p. 931—939 (1987).

78.30. Infrared and Raman spectra and scattering

M. I. Aliev, Kh. A. Khalilov, and G. B. Ibragimov
*Free Carrier Absorption in $In_{1-x}Ga_xAs$
(78.30; S7.12; S7.15)
(b) Vol. **140**, No. 1, p. K83—K86 (1987).

S. Nešpůrek, I. Koropecký, M. Šorm, A. Graja, and E. Sopa
*Physical Properties of the TCNQ Salts with 1-Methyl-3-Alkylimidazolium Derivatives
(72.20 and 78.30; S12)
(b) Vol. **140**, No. 2, p. 545—558 (1987).

V. Železný, J. Petzelt, and R. Swietlik
*Infrared Properties of $TeAE(TCNQ)_2$
(78.30; 63.20; 71.38; S12.1)
(b) Vol. **140**, No. 2, p. 595—604 (1987).

E. Cazzanelli, A. Fontana, F. Rocca, and R. Frech
*Low Temperature Behaviour of the Lithium Modes in Monoclinic Li_2SO_4
(78.30; 63.20; S11.1)
(b) Vol. **140**, No. 2, p. 621—630 (1987).

P. V. Reddy and M. Salagram
*The Far-Infrared Spectra of Mixed Manganese–Magnesium Ferrites
(78.30; S11.2)
(a) Vol. **100**, No. 2, p. 639—643 (1987).

I. Bratu, P. Oană, and M. Culea
*An Infrared Study of $x\%$ BaO · 6 Fe_2O_3 · $(100 - x)\%$ 3 B_2O_3 · 2 BaO Glasses
(78.30; 63.50; S10.15)
(a) Vol. **100**, No. 2, p. K195—K198 (1987).

A. Lehmeyer and L. Merten
Shape of Magnetic Reststrahlen-Bands Caused by Magnon-Polaritons in Uniaxial Ferrites
(71.36 and 75.50; 78.30)
(b) Vol. **141**, No. 1, p. 269—286 (1987).

M. V. Simonyan, E. A. Kafadaryan, M. Ch. Nurijanyan, A. K. Petrosyan, and E. G. Sharoyan
ESR, Optical Absorption and Reflection Spectra of Monoclinic and Triclinic Modifications of Lead Phthalocyanine
(76.30 and 78.20; 78.30; S12.1)
(a) Vol. **101**, No. 1, p. 143—149 (1987).

E. Z. Katsnelson, A. G. Karoza, L. A. Meleshchenko, V. V. Pankov, and B. G. Khavkin
*Detailed Study of IR Absorption Spectra of Manganese–Zinc Ferrites
(78.30; S11.2)
(b) Vol. **141**, No. 2, p. 599—609 (1987).

M. H. Kuok
*Raman Spectra of Cesium Trichlorostannate(II)
(78.30; 63.20; S9)
(b) Vol. **141**, No. 2, p. K167—K171 (1987).

O. A. Golikova and M. M. Kazanin
*Modified Aluminium Dodecaboride
(72.80 and 78.30; S11)
(a) Vol. **103**, No. 1, p. K41—K43 (1987).

A. Förster, S. Kapphan, and M. Wöhlecke
*Overtone Spectroscopy of the OH and OD Stretch Modes in $LiNbO_3$
(78.30 and 78.50; S11.1)
(b) Vol. **143**, No. 2, p. 755—764 (1987).

B. Ulrici, R. Stedman, and W. Ulrici
*Local Vibrational Mode Absorption of Hydrogen and Oxygen Centres in LEC-Grown GaP and GaAs
(63.20 and 78.30; 61.70; S7.11; S7.12)
(b) Vol. **143**, No. 2, p. K135—K139 (1987).

D. A. Jishiashvili and E. R. Kutelia
*Infrared Spectroscopic Study of GeO_x Films
(78.30; S10.1)
(b) Vol. **143**, No. 2, p. K147—K150 (1987).

A. K. Pradhan and R. N. P. Choudhary
*Vibrational Spectra of Rare Earth Orthoniobates
(63.20 and 78.30; S10)
(b) Vol. **143**, No. 2, p. K161—K166 (1987).

A. A. Kaminskii, B. V. Mill, A. V. Butashin, E. L. Belokoneva, and K. Kurbanov
*Germanates with $NdAlGe_2O_7$-Type Structure. Synthesis, Crystal Structure, Absorption–Luminescence Properties, and Stimulated Emission of Their Acivator, Nd^{3+} Ions
(78.30 and 78.55; 61.60; 78.45; S10)
(a) Vol. **103**, No. 2, p. 575—592 (1987).

P. E. Simmonds and A. D. Martin
*Temperature Dependence of Raman Scattering by Optical Phonons in ZnTe
(78.30; S8.13)
(b) Vol. **144**, No. 1, p. K63—K67 (1987).

H. L. Park and W. B. White
*Raman Spectra of FeS and β-CoS
(78.30; S1.63)
(b) Vol. **144**, No. 1, p. K69—K71 (1987).

E. A. Vinogradov, N. M. Gasanly, L. G. Gasanova, A. Z. Magomedov, and V. I. Tagirov
*Infrared Reflectivity Spectra of $Cu_3In_5S_9$ Layer Single Crystals
(78.30; 63.20; S8.16)
(b) Vol. **144**, No. 1, p. K73—K76 (1987).

J. Misiewicz and Z. Gumienny
Impurity Transitions in Zn_3P_2 Observed Using the Wavelength Modulation Technique
(71.55; 78.30; S7)
(a) Vol. **104**, No. 1, p. K37—K40 (1987).

78.35. Brillouin and Rayleigh scattering

78.40. Visible and ultraviolet spectra

DU MAO-LU and ZHAO MIN-GUANG
*A Theoretical Explanation for Pressure Shift of Optical Spectra in Chromium Potassium Alum
(71.70 and 78.40; S11)
(b) Vol. **139**, No. 1, p. K87—K91 (1987).

V. I. GAVRILENKO
Electronic Structure and Optical Properties of Polycrystaalline Cubic Semiconductors
(71.25; 78.40; S5.11)
(b) Vol. **139**, No. 2, p. 457—466 (1987).

K. M. REDDY, A. S. JACOB, B. J. REDDY, and Y. P. REDDY
*Optical Absorption Spectra of Cu^{2+} in Brochantite
(78.30 and 78.40; S11.2)
(b) Vol. **139**, No. 2, p. K145—K150 (1987).

E. SCHULTHEISS, A. SCHARMANN, and D. SCHWABE
*Energy Levels of Triply Ionized Praseodymium in $BiLiF_4$ Crystals
(71.70 and 78.40; 63.20; S9.16)
(b) Vol. **140**, No. 1, p. 173—189 (1987).

J. MÁLEK and M. VANĚČEK
*The Absorption Edge in the n-Type $Ge_{20}Bi_{15}S_{65}$ Glass
(78.40; S41)
(a) Vol. **100**, No. 1, p. K99—K102 (1987).

A. V. SHERMAN
*Energy and Absorption Spectra of One-Dimensional Exciton–Phonon System
(71.35 and 78.40; S12.1)
(b) Vol. **141**, No. 1, p. 151—161 (1987).

J. C. PARLEBAS, K. CHETOUANE, and J. HUGEL
*4f Optical Absorption Model in Rare-Earth Semiconducting Compounds
(71.50 and 78.40; S1.63)
(b) Vol. **142**, No. 1, p. 311—316 (1987).

V. A. ALIEV, E. F. BAGIRZADE, N. Z. GASANOV, and G. D. GUSEINOV
*Behaviour of the Urbach Edge at Phase Transitions in $TlGaSe_2$ Single Crystals
(78.40; 64.70; 71.35; S8.16)
(b) Vol. **143**, No. 2, p. K155—K160 (1987).

78.45. Stimulated emission

K. P. Maheshwari and G. Tarey
*Stimulated Brillouin Scattering of EM Wave Propagating Obliquely in a Magnetoactive Solid State Plasma
(78.35 and 78.45; S7.13)
(b) Vol. **141**, No. 1, p. 333—341 (1987).

A. A. Kaminskii
*On the Laws of Crystal-Field Disorder of Ln^{3+}–Ions in Insulating Crystals
(78.45; 71.70; 78.55; S11)
(a) Vol. **102**, No. 1, p. 389—397 (1987).

A. A. Kaminskii, B. V. Mill, A. V. Butashin, E. L. Belokoneva, and K. Kurbanov
Germanates with $NdAlGe_2O_7$-Type Structure. Synthesis, Crystal Structure, Absorption–Luminescence Properties, and Stimulated Emission of Their Activator, Nd^{3+} Ions
(78.30 and 78.55; 61.60; 78.45; S10)
(a) Vol. **103**, No. 2, p. 575—592 (1987).

78.50. Impurity and defect absorption in solids

J. Lakshmana Rao
Optical Absorption Spectrum of Cr^{3+} Ions Doped in NH_4Cl Single Crystals
(71.55; 78.50; S9)
(b) Vol. **139**, No. 1, p. 241—244 (1987).

Hj. Bernhardt, Th. Dimmig, and A. Dimmig
The Investigation of Point Defects in Lead Molybdate Crystals Which Lead to the Change of Absorption and Fluorescence Properties
(61.70; 78.50; 78.55; S11.1)
(a) Vol. **99**, No. 2, p. 399—404 (1987).

J. C. Krupa, C. Khan Malek, P. Delamoye, B. Moine, and C. Pedrini
*Observation of an Incommensurate Structure by Optical Spectroscopy with a 5f Ion as an Internal Probe. U^{4+}–$ThCl_4$ System
(78.50 and 78.55; 71.70; S9)
(b) Vol. **140**, No. 1, p. 289—300 (1987).

K. Matsuura, S. Kishida, and I. Tsurumi
Annealing of the F^+ EPR Centers and the 2.3 and 2.9 eV Optical Absorption Bands in Zinc Sulfide Crystals
(61.70; 76.30; 78.50; S8.11)
(b) Vol. **140**, No. 2, p. 347—354 (1987).

L. Bosi and M. Nimis
Lifetime Studies on Excited $(F_2^+)^*$ and M Centers in NaF Doped with Magnesium
(61.70; 78.50; S9.11)
(b) Vol. **140**, No. 2, p. 355—360 (1987).

J. Tatarkiewicz, A. Król, A. Breitschwerdt, and M. Cardona
Localized Vibrations of Hydrogen and Deuterium in GaAs. A Comparison with Ge and Si
(63.20; 61.70; 78.50; S7.12)
(b) Vol. **140**, No. 2, p. 369—375 (1987).

W. Ulrici, L. Eaves, K. Kriedland, and D. P. Halliday
Optical Studies of Vanadium in Gallium Phosphide
(71.55; 72.80; 78.50; 78.55; S7.11)
(b) Vol. **141**, No. 1, p. 191—202 (1987).

S. V. Gorbunov, A. V. Kruzhalov, and M. J. Springis
*Optical Properties of the F-Centre in Beryllium Oxide
(78.50 and 78.55; 61.70; 71.55; S10.1)
(b) Vol. **141**, No. 1, p. 293—301 (1987).

A. B. Gerasimov, Z. G. Gogua, and V. N. Svanidze
*Studies of the Optical Impurity-Band Transition Using a Single Potential of an Impurity Centre
(78.50)
(b) Vol. **141**, No. 1, p. K83—K87 (1987).

A. Hamaïdia and J. Margerie
Measurement of Oscillator Strengths of "F_{2A}" Centres in $MeF_2:A^+$ (Me = Ca, Sr; $A^+ = Li^+, Na^+, K^+$)
(61.70; 76.30; 78.50; S9.12)
(b) Vol. **141**, No. 2, p. 391—398 (1987).

M. Herget, A. Hofstaetter, T. Nickel, and A. Scharmann
Reorientation of Hole Centers in $CaWO_4$ Monitored by EPR and Optical Absorption
(71.38 and 71.70; 76.30; 78.50; S11.1)
(b) Vol. **141**, No. 2, p. 523—528 (1987).

S. I. Gaidai, K. P. Meletov, and M. F. Shchanov
Quasi-Resonance in Hydrostatically Compressed Isotopic Mixed Crystals of Deuteronaphthalene
(71.35 and 71.55; 78.50; 78.55; S12)
(b) Vol. **142**, No. 1, p. 211—220 (1987).

K. Matsuura, S. Kishida, I. Tsurumi, and M. Kitagawa
*Electron-Irradiation Effect on the F^+ Bands and the Near-Infrared Emissions in ZnS Crystals
(78.50 and 78.55; 61.70; S8.11)
(b) Vol. **142**, No. 1, p. K79—K83 (1987).

B. Sujak-Cyrul, A. Bartczak, and J. Misiewicz
*The Influence of Heat Treatment on the Infrared Absorption Spectra of Zn_3P_2
(78.50; S7)
(a) Vol. **102**, No. 1, p. 425—432 (1987).

S. Kishida, K. Matsuura, H. Nagase, and I. Tsurumi
*Annealing Characteristics and Temperature Dependence of the Photosensitive 1.9 e.V Absorption Band in Zn-Treated ZnSe Crystals
(78.50; 61.70; S8.12)
(a) Vol. **103**, No. 2, p. 613—618 (1987).

L. Bosi and M. Nimis
Effect of Thermal Expansion on Absorption Band Shift of the F-Center in Alkali Halides
(61.70; 78.50; S9.11)
(b) Vol. **144**, No. 2, p. 501—505 (1987).

T. S. N. Murthy and M. Salagram
Electronic Spectra of Radiation-Induced CrO_4^{3-} in 3 $CdSO_4 \cdot 8\ H_2O$ Crystals
(61.80 and 71.70; 78.50; S11)
(b) Vol. **144**, No. 2, p. 519—527 (1987).

T. F. Veremeichik
*The Regularities in Ground- and Excited-State Absorption Spectra of Impurity Ions of the Iron Group in Crystals
(71.70 and 78.50)
(b) Vol. **144**, No. 2, p. 709—717 (1987).

Yu. E. Perlin, V. M. Polyanovskii, and O. V. Alifanov
*Optical Transitions of Jahn-Teller Local Centres
(71.55 and 78.50)
(b) Vol. **144**, No. 2, p. 817—824 (1987).

M. Okada, T. Kawakubo, T. Seiyama, and M. Nakagawa
*Enhancement of 3d-Electron Transitions in Neutron-Irradiated $MgO:Mn^{2+}$ Crystals
(78.50; 71.70; S10.1)
(b) Vol. **144**, No. 2, p. 903—909 (1987).

E. Radzhabov
*Vacuum Ultraviolet Absorption and Dissociation of H_a^- and OH^- Centers in NaF
(78.50; 61.70; 71.55; S9.11)
(b) Vol. **144**, No. 2, p. K151—K155 (1987).

78.55. Photoluminescence

J. BULLOT and M. P. SCHMIDT
Physics of Amorphous Silicon–Carbon Alloys **(Review Article)**
(61.40; 71.20; 71.55; 72.40; 73.60; 78.20; 78.55; 78.60; S6)
(b) Vol. **143**, No. 2, p. 345—418 (1987).

J. PASTRŇAK, F. KAREL, J. OSWALD, W. ULRICI, and N. M. KOLCHANOVA
The Study of Impurity Levels in LEC Semi-Insulating GaAs:Cr Samples by Two Source Photoconductivity Measurements
(71.55; 72.40; 78.55; S7.12)
(a) Vol. **99**, No. 1, p. 225—235 (1987).

R. STEGMANN, W. N. RASBEGAEV, and A. N. PIKHTIN
*Radiative Recombination in Zn–N-Doped $GaAs_{1-x}P_x$:N Mixed Crystals
(78.55 and 78.60; 71.55; S7.15)
(a) Vol. **99**, No. 1, p. 289—294 (1987).

D. CASSI, R. FAVA, M. MANFREDI, H. OPYRCHAŁ, and M. SUSZYŃSKA
Fractal and Percolative Features of Aggregation Processes in NaCl:Eu^{2+} System Detected by Time Resolved Spectroscopy
(71.55; 78.55; S9.11)
(b) Vol. **139**, No. 2, p. 527—531 (1987).

H. ZIMMERMANN and R. BOYN
Donor-Type Tm Centres in ZnS Crystals
(71.55; 78.55; S8.11)
(b) Vol. **139**, No. 2, p. 533—545 (1987).

T. KESSLER, R. MARKUS, H. NAHME, and N. SCHWENTNER
Nonlinear Excitation of Pure and Doped Rare Gas Crystals
(78.20; 71.35; 78.55; S5.2)
(b) Vol. **139**, No. 2, p. 619—625 (1987).

HJ. BERNHARDT, TH. DIMMIG, and A. DIMMIG
The Investigation of Point Defects in Lead Molybdate Crystals Which Lead to the Change of Absorption and Fluorescence Properties
(61.70; 78.50; 78.55; S11.1)
(a) Vol. **99**, No. 2, p. 399—404 (1987).

H. HAEFNER, M. HEIDER, F. KOCH, R. MITDANK, G. OELGART, W. THIELEMANN, B. RHEINLÄNDER, and B. JACOBS
Properties of Red Light-Emitting (AlGa)As Single-Heterostructure Diodes. I. Structure Characterization
(73.40 and 85; 78.55; 78.60; S7.15)
(a) Vol. **99**, No. 2, p. 657—667 (1987).

R. BOYN and H. ZIMMERMANN
*"Dissolution" of 4f–4f Transitions in Phonon Sidebands (ZnS:Tm^{3+})
(71.55 and 78.55; 71.38; S8.11)
(b) Vol. **140**, No. 1, p. 163—172 (1987).

J. C. KRUPA, C. KHAN MALEK, P. DELAMOYE, B. MOINE, and C. PEDRINI
*Observation of an Incommensurate Structure by Optical Spectroscopy with a 5f Ion as an Internal Probe. U^{4+}–$ThCl_4$ System
(78.50 and 78.55; 71.70; S9)
(b) Vol. **140**, No. 1, p. 289—300 (1987).

W. SIEBERT, R. CARIUS, W. FUHS, and K. JAHN
*Photoluminescence in a-$Si_{1-x}C_x$:H Films
(78.55; 78.65; S6)
(b) Vol. **140**, No. 1, p. 311—321 (1987).

K. LÖBE and R. BOYN
*Dye-Laser Selective 4f–4f Excitation of Electroluminescent Sm^{3+} Centers in ZnS Films and Crystals
(78.55 and 78.60; S8.11)
(b) Vol. **140**, No. 1, p. K91—K95 (1987).

B. RAY, P. A. MONSEF MIRZAI, and J. W. BRIGHTWELL
Fluorescence and ESR in GaS:Mn^{2+}
(71.55; 76.30; 78.55; S8)
(a) Vol. **100**, No. 1, p. 233—237 (1987).

H. HESS
*The Fatiguing of the Photoluminescence in Zinc Silicate Doped with Mn and Ga
(78.55; 72.40; S11.1)
(a) Vol. **100**, No. 1, p. 343—354 (1987).

B. RHEINLÄNDER, W. THIELEMANN, G. OELGART, H. HAEFNER, R. MITDANK, M. HEIDER, F. KOCH, and B. JACOBS
Properties of Red Light-Emitting (Al, Ga)As Single Heterostructure Diodes. II. Luminescence, Injection, and Photoeffect Efficiencies
(73.40 and 85; 78.55; 78.60; S7.15)
(a) Vol. **100**, No. 1, p. 369—377 (1987).

F. LISSE, J. KÖHLER, H. PUFAHL, and D. SCHMID
*Sensitized Luminescence of Pure and Doped $NaNO_2$ Single Crystals. I. Stationary Spectra
(78.55; 71.35; 71.55; S11.1)
(b) Vol. **140**, No. 2, p. 605—619 (1987).

W. ULRICI, L. EAVES, K. FRIEDLAND, and D. P. HALLIDAY
Optical Studies of Vanadium in Gallium Phosphide
(71.55; 72.80; 78.50; 78.55; S7.11)
(b) Vol. **141**, No. 1, p. 191—202 (1987).

S. V. GORBUNOV, A. V. KRUZHALOV, and M. J. SPRINGIS
Optical Properties of the F-Centre in Beryllium Oxide
(78.50; 78.55; 61.70; 71.55; S10.1)
(b) Vol. **141**, No. 1, p. 293—301 (1987).

H. PUFAHL, J. KÖHLER, TH. SCHMIDT, D. SCHMID, and V. M. KENKRE
*Sensitized Luminescence of Pure and Doped $NaNO_2$ Single Crystals. II. Time-Resolved Experiments and Preliminary Analysis
(78.55; 71.35; 71.55; S11.1)
(b) Vol. **141**, No. 1, p. 303—315 (1987).

A. A. Kaminskii
On the Laws of Crystal-Field Disorder of Ln^{3+}–Ions in Insulating Crystals
(78.45; 71.70; 78.55; S11)
(a) Vol. **102**, No. 1, p. 389—397 (1987).

A. V. Vakulenko, R. A. Vardanyan, and L. B. Hovakimyan
*On the Luminescence at the Surface and in the Surface Space Charge Region
(73.20 and 78.55)
(b) Vol. **142**, No. 2, p. 417—422 (1987).

S. N. Konyaev, A. I. Shushin, L. I. Kolesnikova, M. M. Tribel, and E. L. Frankevich
Fluorescence-Detected Low Field Resonances in T–T Annihilation in the A-TCNB Single Crystal. Lineshape and Dynamic Parameters
(71.35; 76.90; 78.55; S12)
(b) Vol. **142**, No. 2, p. 461—471 (1987).

Yu. P. Gnatenko and Yu. I. Zhirko
*Exciton Absorption and Luminescence of Indium Selenide Crystals
(71.35 and 78.55; S8)
(b) Vol. **142**, No. 2, p. 595—604 (1987).

I. Rückmann, M. Petrauskas, V. Netikšis, G. Tamulaitis, and J. Halfpap
*Investigation of p-CdTe and p-$Zn_xCd_{1-x}Te$ Single Crystals by ps-Laser-Induced Gratings and by Photoluminescence
(78.55; S8.13; S8.15)
(b) Vol. **142**, No. 2, p. 629—640 (1987).

P. E. Simmonds and R. Sooryakumar
Stress Effects on Acceptor Bound Excitons in GaAs
(71.35 and 71.55; 78.55; S7.12)
(b) Vol. **142**, No. 2, p. K137—K141 (1987).

W. Hergert, S. Hildebrandt, and L. Pasemann
*Theoretical Investigations of Combined EBIC, LBIC, CL, and PL Experiments. The Information Depth of the PL Signal
(78.55 and 78.60; 72.10)
(a) Vol. **102**, No. 2, p. 819—828 (1987).

K. Matsuura, S. Kishida, Y. Fukata, and I. Tsurumi
*Correlation between the 1.1 and 1.45 eV Emission Bands and the F^+ Optical Absorption Bands in ZnS Crystals
(78.50 and 78.55; 61.70; S8.11)
(b) Vol. **143**, No. 1, p. 275—280 (1987).

A. N. Trukhin
*Temperature Dependence of Luminescence Decay Kinetics of Self-Trapped Excitons, Germanium and Aluminium Centres in Crystalline Quartz
(78.55; 71.35; S10.1)
(b) Vol. **143**, No. 1, p. K83—K88 (1987).

A. K. Uppal, S. N. Chaturvedi, and N. Nath
*Effect of Killer Impurities on the Phosphorescence of CaS:Cu and CaS:Ce
(78.55; S8)
(b) Vol. **143**, No. 1, p. K89—K93 (1987).

J. ZHOU, H. GOTO, N. SAWAKI, and I. AKASAKI
*Near Band Edge Emission of ZnS_xSe_{1-x} Epitaxial Layers on GaP
(78.55 and 78.65; S8.15)
(a) Vol. **103**, No. 2, p. 619—629 (1987).

T. WILSON and P. D. PESTER
Information Depth in Optical Beam Induced Examination of Semiconductor Materials and Devices
(78.20; 78.55)
(a) Vol. **103**, No. 2, p. 631—636 (1987).

M. KITAO, T. NISHIMOTO, and S. YAMADA
*Emission and Excitation Spectra of Photoluminescence of As_2Se_3 Single Crystals
(78.55; S7)
(a) Vol. **103**, No. 2, p. K149—K152 (1987).

I. A. BUYANOVA and S. S. OSTAPENKO
*Excitation Spectra and Polarized Luminescence of Deep Centres in GaAs Single Crystals
(71.55 and 78.55; S7.12)
(a) Vol. **103**, No. 2, p. K161—K164 (1987).

M. S. BRODIN, A. O. GUSHCHA, V. N. KHOTIANTSEV, S. G. SHEVEL, L. V. TARANENKO, and V. V. TISHCHENKO
*Nonlinearities in Excitonic Luminescence of Direct-Gap Semiconductors under Very Low Optical Excitation Levels
(78.55; S8.12; S9)
(b) Vol. **144**, No. 2, p. 863—873 (1987).

A. N. ISOTOV and E. A. STEINMAN
Reconstruction of Optical Dislocation Centres under the Action of Shear Stresses
(61.70 and 71.55; 78.55; S5.11)
(a) Vol. **104**, No. 2, p. 777—784 (1987).

J. WIEHL and S. KEMMLER-SACK
*Cathodo- und Photoluminescence in the Eu^{3+} Activated Garnets of Type $Y_{3-x}Ln_xTe_2Li_3O_{12}$ (Ln = Lu, La) and $Gd_{3-x}La_xTe_2Li_3O_{12}$.
(78.55 and 78.60; 78.30; S11.2)
(a) Vol. **104**, No. 2, p. 863—872 (1987).

F. KOCH and G. OELGART
Reabsorbed Recombination Radiation and Electron Beam Induced Barrier Current in n-Type GaAs
(73.40; 78.20; 78.55; S7.12)
(a) Vol. **104**, No. 2, p. 931—939 (1987).

P. S. GLADKOV, E. N. MONOVA, and K. B. OZANYAN
LPE Growth and Characterization of GaAs:V
(71.55; 72.80; 78.55; S7.12)
(a) Vol. **104**, No. 2, p. K121—K126 (1987).

78.60. Other luminescence and radiative recombination
For photoconduction and photovoltaic effects, see 72.40

K. Löbe and R. Boyn
*Dye-Laser Selective 4f–4f Excitation of Electroluminescent Sm^{3+} Centers in ZnS Films and Crystals
(78.55 and 78.60; S8.11)
(b) Vol. **140**, No. 1, p. K91—K95 (1987).

D. Köhler and E. Kubalek
*Spatially Resolved Cathodoluminescence Investigations on Zinc Oxide Ceramics
(78.60; 61.70; S10.1)
(a) Vol. **100**, No. 1, p. 337—342 (1987).

G. Koschek and E. Kubalek
Scanning Electron Microscope Direct Imaging of Ferroelectric Domains in $BaTiO_3$ Ceramics by Secondary Electrons and Cathodoluminescence
(77.80; 78.60)
(a) Vol. **100**, No. 1, p. 355—368 (1987).

B. Rheinländer, W. Thielemann, G. Oelgart, H. Haefner, R. Mitdank, M. Heider, F. Koch, and B. Jacobs
Properties of Red Light-Emitting (Al, Ga)As Single Heterostructure Diodes. II. Luminescence, Injection, and Photoeffect Efficiencies
(73.40 and 85; 78.55; 78.60; S7.15)
(a) Vol. **100**, No. 1, p. 369—377 (1987).

M. A. S. Sweet and J. Rennie
*Calcium Sulphide as a Thermoluminescence γ-Ray and UV Dosimeter
(78.60; S8)
(a) Vol. **100**, No. 1, p. K67—K71 (1987).

V. Vavilov, I. Gorban, N. Korets, M. Tennakun, I. Tychina, and M. Chukichev
*Cathodoluminescence of Cadmium Diphosphide Crystals
(78.60; 71.55; S7)
(a) Vol. **100**, No. 2, p. 645—651 (1987).

Y. Kirsh, S. Shoval, and P. D. Townsend
*Kinetics and Emission Spectra of Thermoluminescence in the Feldspars Albite and Microcline
(78.60; S11.1)
(a) Vol. **101**, No. 1, p. 253—262 (1987).

P. L. Muthal, B. T. Deshmukh, and S. V. Moharil
*Thermoluminescence in Extended Chain Crystals of Polyethylene
(78.60; S12)
(a) Vol. **101**, No. 2, p. 583—587 (1987).

A. Opanowicz
*On the Kinetics of Thermoluminescence Processes
(78.60; 71.55)
(a) Vol. **101**, No. 2, p. 589—601 (1987).

K. M. Paraskevopoulos, C. Julien, and M. Balkanski
*Fine Structure in the Free Exciton Region of the Emission Spectrum of γ-InSe
(78.60; 71.35; S8)
(b) Vol. **143**, No. 2, p. 741—748 (1987).

H. L. Oczkowski
Kinetic Model with an Intra-Pair Metastable State for Thermally Stimulated Processes
(71.55; 78.60; S8.12)
(a) Vol. **103**, No. 2, p. 499—509 (1987).

R. K. Gartia, S. J. Singh, and A. B. Ahmed
*On the Determination of the Kinetics of Thermally Stimulated Luminescence
(78.60)
(a) Vol. **103**, No. 2, p. 593—598 (1987).

Bhuwan Chandra, S. S. Shinde, A. R. Lakshmanan, and R. C. Bhatt
*Effect of Doping with Monovalent Cation Impurities on the Thermoluminescence Response of $CaSO_4$: Dy Phosphor
(78.60; S11.1)
(a) Vol. **103**, No. 2, p. 599—605 (1987).

W. Hergert and S. Hildebrandt
*On Beam Voltage Dependent Cathodoluminescence Experiments in the Scanning Electron Microscope
(78.60)
(a) Vol. **103**, No. 2, p. K145—K148 (1987).

J. L. Jansons, V. J. Krumins, Z. A. Rachko, and J. A. Valbis
*Luminescence Due to Radiative Transitions between Valence Band and Upper Core Band in Ionic Crystals (Crossluminescence)
(78.60; 71.20; S9.11; S9.12)
(b) Vol. **144**, No. 2, p. 835—844 (1987).

Yu. N. Orlov, V. I. Aleshin, G. E. Arkhangelskii, V. E. Bozhevolnov, Yu. V. Voronov, L. N. Ivanov, V. V. Karelin, and E. I. Saunin
Concentration Peculiarities of a Set of Impurity Centres in Fluorite Crystals Activated by Gadolinium
(61.70; 76.30; 78.60; S9.15)
(b) Vol. **144**, No. 2, p. K77—K80 (1987).

J. Wiehl and S. Kemmler-Sack
*Cathodo- und Photoluminescence in the Eu^{3+} Activated Garnets of Type $Y_{3-x}Ln_xTe_2Li_3O_{12}$ (Ln = Lu, La) and $Gd_{3-x}La_xTe_2Li_3O_{12}$.
(78.55 and 78.60; 78.30; S11.2)
(a) Vol. **104**, No. 2, p. 863—872 (1987).

78.65. Optical properties of thin films

K. Zakrzewska, T. Pisarkiewicz, and A. Czapla
Scattering of Charge Carriers in Reactively Sputtered $CdIn_2O_4$ Thin Films
(73.60; 78.65; S10; S11.2)
(a) Vol. **99**, No. 2, p. 141—149 (1987).

B. P. Rai
*Optical Transmittance of Thin Copper Films in the Visible Region
(78.65; S1.2)
(a) Vol. **99**, No. 1, p. K35—K39 (1987).

G. Jungk and V. Schultze
*Optical Properties of Al Films within and beyond the Effective-Medium Approximation
(78.65; S4)
(b) Vol. **139**, No. 2, p. 627—633 (1987).

M. H. Islam and C. A. Hogarth
*Some Optical and Structural Studies on SiO/SnO_2 Thin Films
(78.65; 68.55; 71.25; S10)
(a) Vol. **99**, No. 2, p. 503—512 (1987).

M. Fathallah, M. Ben Said, and R. Bennaceur
*Determination of the Low Absorption Coefficients of Thin $Cd_{1-x}Zn_xS$ Films by Photothermal Deflection Spectroscopy
(78.65; S8.15)
(a) Vol. **99**, No. 2, p. 521—526 (1987).

W. Siebert, R. Carius, W. Fuhs, and K. Jahn
Photoluminescence in $a\text{-}Si_{1-x}C_x:H$ Films
(78.55; 78.65; S6)
(b) Vol. **140**, No. 1, p. 311—321 (1987).

R. M. Cândea, R. Turcu, G. Borodi, and I. Bratu
Effects of Thermal Annealing in Air on VE, COD, and CAD PbSe Films
(68.55 and 73.60; 78.65; S8)
(a) Vol. **100**, No. 1, p. 149—155 (1987).

A. Kumar, G. Chandra, and O. P. Katyal
*The Effect of Grain Boundary Scattering on the Electrical Resistivity of Polycrystalline Lead, Tin, and Tin–Lead Alloy Films
(72.15 and 73.60; 78.65; S4)
(a) Vol. **100**, No. 1, p. 169—175 (1987).

D. Schalch, A. Scharmann, and R. Wolfrat
*Hydrogenation of Silicon Nitride and Silicon Oxynitride Films Deposited by Reactive Sputtering: Optical Properties
(78.65; S7)
(a) Vol. **100**, No. 1, p. K87—K92 (1987).

R. P. Singh and S. L. Singh
Vacuum-Evaporated Zinc Phosphide Films and Their Characterization
(73.60; 68.55; 78.65; S7)
(a) Vol. **100**, No. 2, p. 493—500 (1987).

P. I. Didenko, A. A. Efremov, V. S. Khomchenko, G. Ph. Romanova, and N. A. Vlasenko
Comparative Study of SiO_x and $SiO_x:TbF_3$ Films
(68.55; 64.75; 78.65; S10.1)
(a) Vol. **100**, No. 2, p. 501—511 (1987).

J. George, B. Pradeep, and K. S. Joseph
*Preparation of Heat Mirrors Using Bismuth Oxide Films
(78.65; 68.55; S10.1)
(a) Vol. **100**, No. 2, p. 513—519 (1987).

N. E. Korsunskaya, I. V. Markevich, and N. R. Stratieva
Photoelectrical and Optical Properties of $Zn_xCd_{1-x}S$ Layers
(72.40 and 73.60; 78.65; S8.15)
(a) Vol. **100**, No. 2, p. 521—525 (1987).

B. Kloth, M. V. Lupal, A. N. Pikhtin, C. E. Richter, R. Ries, R. Stegmann, and M. Trapp
*Determination of the Nitrogen Concentration in VPE-$GaAs_{1-x}P_x$
(68.55 and 78.65; 64.75; S7.15)
(a) Vol. **100**, No. 2, p. 545—552 (1987).

A. K. Abass and M. T. Mohammed
*Investigation of Optically Allowed Transitions of $SnO_2:F$ Coating
(78.65; 71.25; S10.1)
(a) Vol. **100**, No. 2, p. 633—637 (1987).

B. P. Rai
*On the Optical Absorption Edge in Thin GeO_x Films
(78.65; S10.1)
(a) Vol. **100**, No. 2, p. K189—K193 (1987).

L. Wendler and R. Pechstedt
Dynamical Screening, Collective Excitations, and Electron–Phonon Interaction in Heterostructures and Semiconductor Quantum Wells, Application to Double Heterostructures
(71.38 and 71.45; 68.30; 73.40; 77.30; 78.65; S7.12; S7.15)
(b) Vol. **141**, No. 1, p. 129—150 (1987).

C. WÄCHTER and K. HEHL
*General Treatment of Slab Waveguides Including Lossy Materials and Arbitary Refractive Index Profiles
(78.65)
(a) Vol. **102**, No. 2, p. 835—842 (1987).

V. T. BICH, N. N. DINH, T. X. HOAI, N. H. HOANG, L. V. HONG, and V. D. MIEN
Preparation of In_2O_3: Sn (ITO) Films by Electron Beam Deposition
(68.55; 73.60; 78.65; S10)
(a) Vol. **102**, No. 2, p. K91—K94 (1987).

L. WENDLER and R. HAUPT
*An Improved Virtual Mode Theory of ATR Experiments on Surface Polaritons. Application to Long-Range Surface Plasmon-Polaritons in Asymmetric Layer Structures
(68.30 and 78.65; 71.36; 71.45; S1.3)
(b) Vol. **143**, No. 1, p. 131—148 (1987).

A. K. ABASS
*Optical Properties of α-Sulfur Thin Films
(78.65; S5)
(a) Vol. **103**, No. 1, p. 281—284 (1987).

O. GOEDE, W. HEIMBRODT, V. WEINHOLD, E. SCHNÜRER, and H. G. EBERLE
*Optical Study of Phase Transition from Tetrahedrally into Octahedrally Coordinated MnS
(78.65; 68.55; S1.63)
(b) Vol. **143**, No. 2, p. 511—518 (1987).

K. TENNAKONE, A. H. JAYATISSA, C. A. N. FERNANDO, S. WICKRAMANAYAKE, S. PUNCHIHEWA, L. K. WEERASENA, and W. D. R. PREMASIRI
*Semiconducting and Photoelectrochemical Properties of n-and p-Type β-CuCNS
(73. 60 and 78.65; 68.55; 72.40; S11)
(a) Vol. **103**, No. 2, p. 491—497 (1987).

J. GEORGE, B. PRADEEP, and K. S. JOSEPH
*Optical Properties of β-Bi_2O_3 Thin Films
(78.65; S10)
(a) Vol. **103**, No. 2, p. 607—612 (1987).

J. ZHOU, H. GOTO, N. SAWAKI, and I. AKASAKI
*Near Band Edge Emission of ZnS_xSe_{1-x} Epitaxial Layers on GaP
(78.55 and 78.65; S8.15)
(a) Vol. **103**, No. 2, p. 619—629 (1987).

S. A. EL-SAHHAR, M. EL-SHABASY, and H. M. TALAAT
*Study of the Structure and Optical Properties of Thin Zinc Films
(68.55 and 78.65; S1.2)
(a) Vol. **104**, No. 2, p. 755—760 (1987).

78.70. Other interactions of matter with particles and radiation

N. M. KABACHNIK, H. KERKOW, V. P. PETUKHOV, E. A. ROMNOVSKII, and R. WEDELL
*Ion-Reduced Characteristic X-Ray Emission in Solids **(Review Article)**
(78.70; 61.80)
(b) Vol. **139**, No. 1, p. 17—57 (1987).

A. L. IVANOVSKII, D. L. NOVIKOV, and V. A. GUBANOV
Electronic Properties of Carbides, Nitrides, and Oxides of Subgroup VIa Transition Metals. Theory and Experiment **(Review Article)**
(71.25; 73.20; 78.70; 79.60; S1.61; S1.62; S10)
(b) Vol. **141**, No. 1, p. 9—33 (1987).

G. DLUBEK and R. KRAUSE
*Positron Studies of Defects in III-V Semiconductor Compounds **(Review Article)**
(61.70 and 78.70; S7.1)
(a) Vol. **102**, No. 2, p. 443—479 (1987).

H. P. BONDE and S. B. SHRIVASTAVA
Electron Momentum Density Distributions in Rhodium
(71.25; 78.70; S1; S1.3)
(b) Vol. **139**, No. 1, p. 155—160 (1987).

R. UBGADE and P. R. SARODE
*Study of Strontium Compounds and Minerals by X-Ray Absorption Spectroscopy. I. K-Edge Shifts
(78.70; S3; S9.12; S10; S11.1)
(a) Vol. **99**, No. 1, p. 295—301 (1987).

M. MOUSSAVI-MADANI and P. C. RICE-EVANS
*Search for Positron Trapping in Trigonal Selenium
(61.70 and 78.70; S5)
(b) Vol. **139**, No. 2, p. K99—K102 (1987).

A. URBANIAK-KUCHARCZYK
*Spin Autocorrelation Time in Thin Iron, Nickel, and Gadolinium Films
(78.70; S1.1; S1.2; S1.4)
(b) Vol. **139**, No. 2, p. K141—K143 (1987).

S. SZUSKIEWICZ
A Study of the Electronic Structure of $Fe_{1-x}Co_x$ Alloys by a Polarized Positron Annihilation
(71.25; 78.70; S1.1; S1.2)
(b) Vol. **140**, No. 1, p. 141—147 (1987).

H. STACHOWIAK
Screening of a Proton Moving through the Electron Gas. II. Kinetic Properties of the Screening Cloud
(71.25; 78.70)
(a) Vol. **140**, No. 2, p. 521—526 (1987).

C. Sugiura, M. Kitamura, and S. Muramatsu
*Ruthenium L_{II} and Rhodium $L_{II,III}$ X-Ray Absorption-Edge Structures
(78.70; S9.16)
(b) Vol. **140**, No. 2, p. 631—636 (1987).

B. K. Sharma, Hanuman Singh, S. Perkkiö, T. Paakkari, and K. Mansikka
Electron Momentum Distribution in Silver and the Renormalized Free Atom Model
(71.25; 78.70; S1.3)
(b) Vol. **141**, No. 1, p. 177—184 (1987).

M. Szuszkiewicz, S. Szuszkiewicz, and J. Chojcan
*Back-Scattering of Positrons on Magnetized Samples
(78.70; S1.1; S1.2)
(b) Vol. **142**, No. 1, p. K79—K81 (1987).

S. Kopta, R. Hajduk, A. Z. Hrynkiewicz, S. Łazarski, E. Maydell-Ondrusz, and B. Rajchel
Rutherford Backscattering, Coulomb Excitation, and X-Ray Production Studied in Channeling Conditions
(61.80; 78.70; S1)
(a) Vol. **101**, No. 1, p. 77—96 (1987).

K. Hemachandran, A. R. Chetal, and G. Joshi
X-Ray Absorption Spectroscopic Studies of Chalcopyrite Mineral
(71.25; 78.70; S1.63)
(b) Vol. **141**, No. 2, p. 441—445 (1987).

H. Stachowiak
Screening of a Proton Moving through the Electron Gas. III. Computation of the Momentum Distribution of the Screening Cloud
(71.25; 78.70)
(b) Vol. **141**, No. 2, p. 447—456 (1987).

C. Sugiura, M. Kitamura, and S. Muramatsu
*Palladium L_{III} X-Ray Absorption-Edge Spectra of $Pd(ac)_2$, $Pd(acac)_2$, and $Pd_2(dba)_3 \cdot CHCl_3$
(78.70; S12)
(b) Vol. **141**, No. 2, p. K173—K177 (1987).

V. Fritzsche and P. Rennert
*A Curved-Wave Theory for EXAFS with Effective Scattering Amplitudes
(61.10 and 78.70; S1.2)
(b) Vol. **142**, No. 1, p. 15—25 (1987).

V. I. Kruglov
*Interaction of Semiconductors with a Strong Resonance Electromagnetic Field
(71.45 and 78.70)
(b) Vol. **142**, No. 1, p. 165—178 (1987).

M. Kitamura, S. Muramatsu, and C. Sugiura
*Multiple Scattering Study for Cl K XANES of KCl, $SrCl_2$, and CsCl Crystals
(71.10 and 78.70; S9.11; S9.12)
(b) Vol. **142**, No. 1, p. 191—202 (1987).

M. ELDRUP and K. O. JENSEN
*Positron Trapping Rates into Cavities in Al: Temperature and Size Effects
(61.70 and 78.70; S4)
(a) Vol. **102**, No. 1, p. 145—152 (1987).

K. GHAZI-WAKILI, U. ZIMMERMANN, J. BRUNNER, P. TIPPING, W. B. WAEBER, and F. HEINRICH
*Positron Annihilation Studies on Neutron Irradiated Pressure Vessel Steels
(61.80 and 78.70; 62.20; 64.75; S1.1)
(a) Vol. **102**, No. 1, p. 153—163 (1987).

ZHU GUANG-QU, LU SHAN-YI, and WANG SHAO-JIE
*Investigation of Microstructure and Coercivity Mechanisms in the Permanent Magnet Alloy Mn–Al–C by Positron Annihilation and Transmission Electron Microscopy
(75.60 and 78.70; S1.2)
(a) Vol. **102**, No. 1, p. 165—169 (1987).

S. DANNEFAER
*Defects and Oxygen in Silicon Studied by Positrons
(61.70 and 78.70; S5.11)
(a) Vol. **102**, No. 2, p. 481—491 (1987).

J. C. BOURGOIN, H. J. VON BARDELEBEN, and D. STIEVENARD
Irradiation Induced Defects in III–V Semiconductor Compounds
(61.80; 61.70; 78.70; S7.1)
(a) Vol. **102**, No. 2, p. 499—510 (1987).

T. GOWOREK
*Defects in Molecular Solids. Positron Studies
(61.70 and 78.70; S12)
(a) Vol. **102**, No. 2, p. 511—526 (1987).

W. PUFF, S. DANNEFAER, P. MASCHER, and D. KERR
*Defect Characteristics in Different Crystallographic Directions in Cz-Si as a Function of Doping and Annealing
(61.70 and 78.70; S5.11)
(a) Vol. **102**, No. 2, p. 527—531 (1987).

BIZHONG ZHOU, JIANGLING FONG, and JINGZHAO HUANG
*Investigation of Defects in GaP Using Positron Annihilation
(61.70 and 78.70; S7.11)
(a) Vol. **102**, No. 2, p. 533—536 (1987).

C. D. BELING, R. I. SIMPSON, M. G. STEWART, Y. Y. WANG, S. FUNG, J. C. H. WAI, and T. N. SUN
*Positron Mobility Measurements and Their Relevance to Defect and Impurity Studies in Semiconductors and Insulators
(61.70 and 78.70; S5.11; S7.11)
(a) Vol. **102**, No. 2, p. 537—547 (1987).

D. SCHÖDLBAUER, G. KÖGEL, P. SPERR, and W. TRIFTSHÄUSER
*Lifetime Measurements with a Pulsed Slow Positron Beam
(78.70; S1.1; S1.2)
(a) Vol. **102**, No. 2, p. 549—554 (1987).

D. E. Popov, E. I. Rozum, S. E. Savushkin, S. R. Uglov, and S. A. Vorobiev
*Optical Radiation of Electrons and Positrons at Quasi-Molecular Channeling in Crystals
(71.25 and 78.70)
(b) Vol. **143**, No. 2, p. K141—K145 (1987).

Cs. Szeles, K. Süvegh, Z. Homonnay, and A. Vértes
Positron Lifetime and Mössbauer Spectroscopy Study of Vacancy–Tin Interaction in Dilute Al–Sn Alloys
(61.70; 76.80; 78.70; S4)
(a) Vol. **103**, No. 2, p. 397—401 (1987).

I. Kanazawa, H. Murakami, and M. Doyama
The Migration Stage of Divacancies in Quenched Pure Aluminium, Dilute Aluminium Alloys, and Pure Copper Studied by Positron Annihilation
(61.70; 78.70; S1.2; S4)
(a) Vol. **103**, No. 2, p. 403—411 (1987).

S. Szuszkiewicz
*A Determination of the Polarization of Conduction Electrons by Polarized Positron Annihilation
(75.25 and 78.70)
(b) Vol. **144**, No. 1, p. K45—K50 (1987).

H.-E. Schaefer and F. Banhart
Thermal Equilibrium Vacancies in Platinum Studied by Positron Annihilation
(61.70; 71.60; 78.70; S1.3)
(a) Vol. **104**, No. 1, p. 263—272 (1987).

P. Kindl and G. Reiter
Investigations on the Low-Temperature Transitions and Time Effects of Branched Polyethylene by the Positron Lifetime Technique
(64.60; 78.70; S12)
(a) Vol. **104**, No. 2, p. 707—713 (1987).

P. Pająk and B. Rozenfeld
*Helium in Plastically Deformed Ni Samples Investigated by Means of the Positron Annihilation Method
(61.70 and 78.70; S1.2)
(a) Vol. **104**, No. 2, p. K85—K88 (1987).

79. Electron and ion emission by liquids and solids; impact phenomena

79.20. Impact phenomena, including electron spectra and sputtering

A. P. Dmitriev, M. P. Mikhailova, and I. N. Yassievich
*Impact Ionization in $A^{III}B^{V}$ Semiconductors in High Electric Fields **(Review Article)**
(79.20; S7.1)
(b) Vol. **140**, No. 1, p. 9—37 (1987).

D. Lovrić and B. Gumhalter
Physisorption and Chemisorption Effects of Atom–Adsorbate van der Waals Potentials. Application to He and He* Scattering from Adsorbed CO
(68.20 and 73.90; 79.20)
(b) Vol. **139**, No. 2, p. 423—432 (1987).

P. Godowski and J. Rudny
Effect of Sulphur on the Surface Segregation in $Co_{50}Ni_{50}$ Alloy
(68.20; 79.20; S1.2)
(a) Vol. **99**, No. 2, p. 527—536 (1987).

M. Procop, H. Raidt, and B. Sandow
Auger Electron and X-Ray Spectroscopic Analysis of Au-Ge Contacts to GaAs
(68.48; 73.40; 79.20; S1.3; S7.12)
(a) Vol. **99**, No. 2, p. 573—580 (1987).

W. Th. M. Buters, A. van Ven, and A. van den Beukel
Some Results on Helium Trapping in Undeformed and Cold Worked Single Crystalline Copper. A Comparison with Molybdenum and Nickel
(61.70; 68.45; 79.20; S1.2)
(a) Vol. **100**, No. 1, p. 87—94 (1987).

T. Jung and M. Schmidt
Plasma-Assisted Reactive Evaporation of Aluminium Nitride Films
(68.55 and 73.60; 79.20; S7.1)
(a) Vol. **100**, No. 1, p. 207—211 (1987).

M. Hacskaylo
A Frequency Shift of Light at an Optical Boundary
(78.20; 79.20)
(a) Vol. **100**, No. 2, p. 653—657 (1987).

L. Guoba, G. Zubauskas, J. Kameneckas, and A. Sakalas
Study of Multilayer Structures on GaAs by Rutherford Backscattering with Carbon Ions
(68.48; 79.20; S7.12)
(a) Vol. **101**, No. 2, p. 445—450 (1987).

Deyan He, Fangqing Zhang, Jingzhong Cui, and Guanghua Chen
Interface Defects in Sputtered a-Si: H/a-C: H Multilayers
(73.40; 76.30; 79.20; S5; S5.11)
(b) Vol. **142**, No. 1, p. K35—K37 (1987).

S. P. Chenakin, V. T. Cherepin, A. L. Pivovarov, and M. A. Vasilev
*Temperature Dependence of Secondary Ion Emission from Iron-Based Crystalline and Amorphous Alloys
(61.55 and 79.20; S1.1)
(a) Vol. **102**, No. 1, p. 231—240 (1987).

E. Taute
*Auger Neutralization of Fast Ions at Surfaces
(79.20)
(b) Vol. **142**, No. 2, p. 437—443 (1987).

B. Rauschenbach
The Characteristic Temperature of Ion Beam Mixing
(61.80; 79.20; S1; S4)
(a) Vol. **102**, No. 2, p. 645—651 (1987).

H. Boudriot, K. Deus, R. Gründler, H. A. Schneider, and T. Stöckert
*Optical Properties of $ZnSiAs_2$ Single Crystalline Thin Films from Electron Energy Loss Measurements
(78.20 and 79.20; 71.25; S7.16)
(b) Vol. **143**, No. 2, p. K125—K130 (1987).

R. C. Martin and N. M. Ghoniem
Monte Carlo Simulation of Coupled Ion–Electron Transport in Semiconductors
(61.80 and 73.40; 79.20; S5.11)
(a) Vol. **104**, No. 2, p. 743—754 (1987).

M. Procop, B. Sandow, H. Raidt, and Li Do Son
The Importance of the Ni to Ge Ratio and of the Annealing Cycle for the Resistivity and Morphology of NiAuGe Ohmic Contacts to n-GaAs
(73.40 and 85; 68.48; 79.20; S7.12)
(a) Vol. **104**, No. 2, p. 903—916 (1987).

H. Johansen and B. Lämmel
Optical Feedback during CW Laser Annealing of Silicon Detected by SEM-EBIC
(61.80; 79.20; S5.11)
(a) Vol. **104**, No. 2, p. K83—K84 (1987).

79.40. Thermionic emission

M. Parlapanski
*Thermionic Emission from Irregular Metal Surfaces
(68.20 and 79.40; S2)
(b) Vol. **139**, No. 1, p. K5—K8 (1987).

K. Biedrzycki and B. Kosturek
*Thermally Stimulated Electron Emission from Triglycine Sulphate Crystals with Well-Defined Initial Domain Structure
(77.80 and 79.40; S12)
(a) Vol. **100**, No. 1, p. 327—330 (1987).

G. Jungk
*Is the 1.5 eV Interband Transition in Al Films an Experimental Illusion?
(79.40; S4)
(b) Vol. **141**, No. 1, p. K75—K78 (1987).

79.60. Photoemission and photoelectron spectra

A. L. Ivanovskii, D. L. Novikov, and V. A. Gubanov
Electronic Properties of Carbides, Nitrides, and Oxides of Subgroup VIa Transition Metals. Theory and Experiment **(Review Article)**
(71.25; 73.20; 78.70; 79.60; S1.61; S1.62; S10)
(b) Vol. **141**, No. 1, p. 9—33 (1987).

A. A. Karpushin, A. N. Sorokin, I. Yu. Semenova, and M. Tomášek
A Contribution to the Theoretical Interpretation of the UPS Spectra of Monoatomic Adsorption of Oxygen on the Si (111) Surface
(68.20 and 73.20; 79.60; S5.11)
(b) Vol. **139**, No. 1, p. 125—130 (1987).

J. Mašek and B. Velický
Mn 3d States in Photoelectron Spectra from $Cd_{1-x}Mn_xTe$
(71.25; 79.60; S8.15)
(b) Vol. **140**, No. 1, p. 135—140 (1987).

C. Majumdar, A. B. Maity, and A. N. Chakravarti
*Effect of Size Quantization on Photoemission from Ultrathin Films of Wide-Gap Semiconductors
(73.60 and 79.60; S7.12)
(b) Vol. **140**, No. 1, p. K7—K12 (1987).

B. ROTHER, C. WEISSMANTEL, and G. LEONHARDT
X-Ray Photoelectron Spectroscopy at Ion-Plated Boron Nitride
(68.55; 79.60; S7)
(a) Vol. **100**, No. 2, p. 553—557 (1987).

C. MAJUMDAR, A. B. MAITY, and A. N. CHAKRAVARTI
*On the Photoemission from Quasi-One-Dimensional Structures of Wide-Gap Semiconductors Having Rectangular Cross-Sections
(73.60 and 79.60)
(b) Vol. **141**, No. 1, p. K35—K40 (1987).

N. WAGNER, M. HORX, M. JACOBS, and M. SUCHI
*XPS Investigations of TiO_2/RuO_2 Mixed Oxide Films on Titanium
(68.48 and 79.60; S10.15)
(a) Vol. **101**, No. 1, p. K23—K27 (1987).

R. MAMY, A. BOUFELJA, and B. CARRICABURU
Angle Resolved Photoemission and Electronic Band Structure of MoS_2
(71.25; 79.60; S8)
(b) Vol. **141**, No. 2, p. 467—473 (1987).

A. CHASSÉ and P. RENNERT
*Theoretical Angle-Resolved X-Ray Photoelectron Spectra from GaAs (110)
(79.60; S7.12)
(b) Vol. **141**, No. 2, p. K147—K152 (1987).

YU. D. DEKHTYAR and G. L. SAGALOVICH
*Photothermostimulated Exoelectron Emission from Crystalline Silicon Containing Edge Dislocations and Point Defects
(79.60; 61.70; S5.11)
(a) Vol. **102**, No. 2, p. 665—677 (1987).

A. BECKMANN
*On the Theory of Photoemission from Adsorbate-Induced Surface States
(73.20 and 79.60; S1.2)
(b) Vol. **144**, No. 1, p. 213—224 (1987).

C. MAJUMDAR, A. B. MAITY, and A. N. CHAKRAVARTI
*Quantized Photoemission from Quasi-Two-Dimensional Structures of Wide-Gap Semiconductors in a Strong Magnetic Field
(73.60 and 79.60; S7.12)
(b) Vol. **144**, No. 1, p. K13—K18 (1987).

J. BRAUN, G. THÖRNER, and G. BORSTEL
*Relativistic Theory of Photoemission from Solids. III. Bulk and Surface Transition Matrix Elements
(79.60; 73.20)
(b) Vol. **144**, No. 2, p. 609—616 (1987).

A. B. MAITY, C. MAJUMDAR, and A. N. CHAKRAVARTI
*On Photoemission from Quasi-One-Dimensional Structures of Narrow-Gap Semiconductors Having Rectangular Cross-Sections
(73.20 and 79.60; S7.13)
(b) Vol. **144**, No. 2, p. K93—K98 (1987).

79.70. Field emission and field ionization

C. Majumdar, M. K. Bose, A. B. Maity, and A. N. Chakravarti
Effect of Size Quantization on Field Emission from Ultrathin Films of Degenerate Wide-Gap Semiconductors
(73.60; 79.70; S7.12)
(b) Vol. **141**, No. 2, p. 435—439 (1987).

M. K. Bose, C. Majumdar, A. B. Maity, and A. N. Chakravarti
*On Field Emission from Ultrathin Films of Semiconductors with Effects of Image Forces
(73.60 and 79.70; S7.12)
(b) Vol. **143**, No. 1, p. 113—120 (1987).

79.75. Exoelectric emission

E. Chryssou
*Thermally Stimulated Exoelectron Emission (TSEE) of Heat Treated K_2SO_4 and of Sm-Dy-Doped K_2SO_4
(79.75; S11.1)
(a) Vol. **99**, No. 2, p. K103—K108 (1987).

Y. Fukuda, A. Tomita, and N. Takeuchi
*Thermoluminescence and Thermally Stimulated Exoelectron Emission of Sintered CaB_4O_7 Doped with Pb, Eu, or Dy
(78.60 and 79.75; S10)
(a) Vol. **99**, No. 2, p. K135—K138 (1987).

M. M. Merkin and A. A. Tsertsvadze
*Exoemission Theory: The Exciton Model
(79.75; 61.70; 71.35; S9.11)
(b) Vol. **140**, No. 1, p. 73—82 (1987).

E. Chryssou and P. Vartzeli-Nikaki
*Exoelectron Emission (EE) of Heat Induced $SrSO_4$/SrS Conversion
(73.20 and 79.75; S11.1)
(a) Vol. **104**, No. 2, p. K111—K116 (1987).

79.80. Resonance tunneling

H. Yamamoto
*Resonant Tunneling with the Mass Variation through Rectangular Double-Barrier Structure
(79.80)
(b) Vol. **140**, No. 1, p. K23—K25 (1987).

85. Devices

A. SRIVASTAVA
*Threshold Voltage Dependence on Channel Length in Small Geometry MOSFETs
(73.40 and 85)
(a) Vol. **99**, No. 1, p. 303—308 (1987).

H. K. PHAN, L. H. PHU, and N. T. NGUYET ANH
*Switching Characteristics of MISS Devices Containing Electron Surface Traps
(73.40 and 85; S5.11)
(a) Vol. **99**, No. 1, p. 309—314 (1987).

B. RHEINLÄNDER, J. KOVÁC, P. NEVERMANN, and H. FIESELER
*On the Determination of Quantum Yield in InGaAsP Double-Heterostructure LED
(73.40 and 85, S7.15)
(a) Vol. **99**, No. 1, p. 315—320 (1987).

M. ZAFAR IQBAL, A. JABBAR, N. BABER, and N. ZAFAR
*An Inadvertent Mid-Gap Electron Level in Liquid-Phase-Epitaxial GaP
(71.55 and 85; S7.11)
(a) Vol. **99**, No. 1, p. K65—K68 (1987).

M. KITTLER and W. SEIFERT
Characterization of the Leakage-Current Diffusion Component in Intrinsically Gettered Silicon by EBIC
(73.40; 72.40; 85; S5.11)
(a) Vol. **99**, No. 2, p. 559—572 (1987).

F. JANUSCHEWSKI and H.-J. ERZGRÄBER
*Modelling the Influence of Hot Electrons on the Transfer Characteristic of Short-Channel MOSFETs
(73.40 and 85)
(a) Vol. **99**, No. 2, p. 649—656 (1987).

H. HAEFNER, M. HEIDER, F. KOCH, R. MITDANK, G. OELGART, W. THIELEMANN, B. RHEINLÄNDER, and B. JACOBS
*Properties of Red Light-Emitting (AlGa)As Single-Heterostructure Diodes. I. Structure Characterization
(73.40 and 85; 78.55; 78.60; S7.15)
(a) Vol. **99**, No. 2, p. 657—667 (1987).

N. T. TRAN and F. R. JEFFREY
*Effects of Carbon Profiles at the p/i Interface in Amorphous Silicon Solar Cells
(73.40 and 85; S5.11; S6)
(a) Vol. **99**, No. 2, p. K143—K148 (1987).

G. GHIBAUDO
*A Simple Model of the Drain Saturation Voltage Dependence with Gate Voltage for Short-Channel MOSFETs
(73.40 and 85)
(a) Vol. **99**, No. 2, p. K149—K153 (1987).

P. Singh and B. Baishya
*Some Studies on Thin Film Transistors (TFTs) with Er_2O_3 as Gate Insulator
(73.40 and 85; S8.11; S8.12)
(a) Vol. **101**, No. 2, p. K173—K177 (1987).

G. J. Papaioannou
*Electron Radiation Effects on GaAs MESFETs
(61.80 and 85; 73.40; S7.12)
(a) Vol. **102**, No. 2, p. 843—848 (1987).

A. I. Kasiyan, A. G. Cheban, M. A. Cheban, I. V. Sur, and O. V. Zakharchuk
*Effect of Traps on the Shape of the Output Signal of Ionizing-Radiation Semiconductor Detectors
(72.20 and 85; S5.11)
(a) Vol. **103**, No. 1, p. 185—191 (1987).

G. Ghibaudo
*An Analytical Model of Conductance and Transconductance for Depletion-Mode MOS Transistors
(73.40 and 85)
(a) Vol. **103**, No. 1, p. 317—325 (1987).

P. Singh and B. Baishya
*Suitability of Rare-Earth Oxides for Use in Thin Film Transistors
(73.40 and 85; S10)
(a) Vol. **104**, No. 2, p. 885—889 (1987).

M. Procop, B. Sandow, H. Raidt, and Li Do Son
*The Importance of the Ni to Ge Ratio and of the Annealing Cycle for the Resistivity and Morphology of NiAuGe-Ohmic Contacts to n-GaAs
(73.40 and 85; 68.48; 79.20; S7.12)
(a) Vol. **104**, No. 2, p. 903—916 (1987).

G. J. Papaioannou, C. D. Kourkoutas, N. G. Theofanous, and A. T. Arapoyianni
*A Study of the Gain Mechanisms in GaAs Planar Optical Photoconductive Detectors
(72.40 and 85; S7.12)
(a) Vol. **104**, No. 2, p. K141—K146 (1987).

S1. Transition metals and their alloys and compounds
(for oxides, see S10)

H.-E. SCHAEFER
Investigation of Thermal Equilibrium Vacancies in Metals by Positron Annihilation
(61.70; 71.60; 78.70; S1; S4)
(a) Vol. **102**, No. 1, p. 47—65 (1987).

Zs. KAJCSOS
Properties of Amorphous and Fine-Grained Alloys as Seen by Positron Annihilation
(61.40 and 78.70; S1)
(a) Vol. **102**, No. 1, p. 67—78 (1987).

C. HIDALGO, S. LINDEROTH, and N. DE DIEGO
Positron–Dislocation Interaction in Deformed Zn and Cd
(71.60; 61.70; S1; S1.2)
(a) Vol. **102**, No. 1, p. 113—117 (1987).

A. BELAIDI, H. P. LEIGHLY, P. G. COLEMAN, and R. N. WEST
A Positron Study of Defect Formation in Cd–Mg Alloys
(64.75 and 78.70; 61.70; S1)
(a) Vol. **102**, No. 1, p. 127—132 (1987).

V. E. ANTONOV, I. T. BELASH, O. V. ZHARIKOV, and A. V. PALNICHENKO
A Study on Superconductivities of High-Pressure Phases in the Re–H, Ru–H, Rh–H, and Ni–H Systems
(74.10; S1)
(b) Vol. **142**, No. 2, p. K155—K159 (1987).

B. RAUSCHENBACH
The Characteristic Temperature of Ion Beam Mixing
(61.80; 79.20; S1; S4)
(a) Vol. **102**, No. 2, p. 645—651 (1987).

S. NAGATA, M. OGINO, and S. TANIGUCHI
Electrical Resistivity of Thin Metal Films Vapor-Quenched at 77 K. Cu, Ag, Au, Ni, Pd, and Pt
(73.60; S1; S1.2; S1.3)
(a) Vol. **102**, No. 2, p. 711—717 (1987).

C. SCHOBER and V. N. ANTONOV
The Gyromagnetic Factor of Conduction Electrons in Noble and Transition Metals
(71.25; S1; S1.3)
(b) Vol. **143**, No. 1, p. K31—K35 (1987).

S. FUJITA
Kinetic Theory of the Hydrogen Diffusion in Metals
(66.30; S1)
(b) Vol. **143**, No. 2, p. 443—451 (1987).

J. SOSNOWSKI
Superconducting and Magnetic Properties of A15-Type Nb_3Ga Alloys Doped with 3d Impurities
(74.70; 75.50; S1)
(a) Vol. **103**, No. 2, p. 531—542 (1987).

S1.1. Fe

A. Ostrasz and M. Szuskiewicz
3d-Like Electron Localization and Charge Transfer in Hydrogenated V–Fe Alloys
(71.25; 61.55; 76.80; S1.1; S1.2)
(b) Vol. **139**, No. 1, p. 205—211 (1987).

I. Retat
Plastic Deformation of Fe–Ni and $Fe_{65}(Ni_{1-x}Mn_x)_{35}$ between 30 and 460 K. Magnetically Induced Invar Hardening
(62.20; 75.80; S1.1; S1.2)
(a) Vol. **99**, No. 1, p. 121—130 (1987).

J. M. Barandiarán, J. C. Gómez-Sal, J. Rodríguez Fernández, R. J. López Sánchez and O. V. Nielsen
Temperature Dependence of the Electrical Resistivity in $[Co_{1-x}(Fe_{0.5}Ni_{0.5})_x]_{75}Si_{15}B_{10}$ Metallic Glasses
(72.15; S1.1; S1.2)
(a) Vol. **99**, No. 1, p. 243—249 (1987).

I. Okamoto, T. Miyazaki, and M. Takahashi
Change of Magnetic Properties with Heat Treatments in Rapidly Quenched $Cu_{99.2}Fe_{0.8}$ Alloys
(75.20 and 75.50; S1.1; S1.2)
(a) Vol. **99**, No. 1, p. 267—274 (1987).

I. Škorvánek and A. Zentko
Effects of Neutron Irradiation on Magnetic Properties of $Fe_{80}B_{20}$ Amorphous Alloys
(75.50; 61.80; 75.30; 75.60; S1.1)
(a) Vol. **99**, No. 1, p. 275—278 (1987).

R. A. Petrova
Kinetics of Formation of the σ-Phase in Austenit-Ferrite Stainless Steel Type X18AΓ12
(61.55 and 64.70; S1.1)
(a) Vol. **99**, No. 1, p. K1—K5 (1987).

J. Li and B. X. Liu
CEMS to Characterize the Ion Induced Fe–Mo Metastable Phase
(64.60 and 76.80; S1; S1.1)
(a) Vol. **99**, No. 1, p. K31—K34 (1987).

K.-H. Müller, A. Handstein, D. Eckert, and J. Schneider
The Dip in Magnetization Curves of Sintered Nd–Fe–B Permanent Magnets
(75.60; S1.1; S1.4)
(a) Vol. **99**, No. 1, p. K61—K64 (1987).

K. Terao and M. Shimizu
Electronic Structure and Magnetism of AFe_2 Compounds with C14 and C15 Laves Structures
(71.25 and 75.50; 75.80; S1.1)
(b) Vol. **139**, No. 2, p. 485—497 (1987).

V. V. Kolomytkin
The Pipe Diffusion of the Self-Interstitial in a Loaded Crystal
(66.30; S1.1)
(a) Vol. **100**, No. 1, p. K17—K20 (1987).

O. Kohmoto and N. Yamaguchi
Shift of Peak Position in μ–T Curves for Amorphous FeCo–SiB Alloys Having Zero Magnetostriction
(75.50 and 75.80; S1.1; S1.2)
(a) Vol. **100**, No. 1, p. K61—K65 (1987).

H. Hermann, N. Mattern, and W. Matz
Magnetic and Structural Properties of Rapidly Quenched Iron–Boron Metallic Glasses
(61.40 and 75.50; S1.1)
(b) Vol. **140**, No. 2, p. 581—588 (1987).

M. Budzyński, O. I. Kochetov, M. Subotowicz, H. Niezgoda, H. Spustek, and W. Tańska-Krupa
Influence of Hydrogenation on the Hyperfine Interactions in $(Y_{0.97}Hf_{0.03})Fe_2$
(75.20; 75.25; 75.50; 76.80; S1.1; S1.4)
(b) Vol. **140**, No. 2, p. 589—594 (1987).

B. Rauschenbach and V. Heera
Ion Irradiation Induced Reamorphization of Boron Preimplanted Iron
(61.40 and 61.80; S1.1)
(a) Vol. **100**, No. 2, p. 423—429 (1987).

H. Schäfer and H. H. Mende
Stray Field Profile of Grain-Oriented Rectangular 3% SiFe Sheets in Dependence on Vertical and In-Plane Magnetic Fields
(75.50; S1.1)
(a) Vol. **100**, No. 2, p. 621—632 (1987).

P. Kopčanský
The Electrical Resistivity of the Amorphous Fe–Ni–B System after Neutron Irradiation — Model of a Spinodal Decomposition
(72.15; S1.1)
(a) Vol. **100**, No. 2, p. K159—K163 (1987).

A. Węgrzyn, L. T. Baczewski, and Z. Obuszko
Magnetization and Magnetostriction of Iron-Rich $(Fe_{1-x}Co_x)_{90}Zr_{10}$ Metallic Glasses
(75.80; S1.1; S1.2)
(a) Vol. **100**, No. 2, p. K181—K184 (1987).

R. Krishnan and M. Dancygier
High Field Susceptibility Peak in Some NiFeSiB Amorphous Alloys
(75.60; S1.1; S1.2)
(b) Vol. **141**, No. 1, p. K71—K74 (1987).

M. Szuszkiewicz, S. Szuszkiewicz, and J. Chojcan
Back-Scattering of Positrons on Magnetized Samples
(78.70; S1.1; S1.2)
(b) Vol. **141**, No. 1, p. K79—K81 (1987).

L. Dobrzyński, T. Giebułtowicz, M. Kopcewicz, M. Piotrowski, and K. Szymański
Neutron and Mössbauer Studies of $Fe_{3-x}Al_xSi$ Alloys
(61.55 and 76.80; S1.1)
(a) Vol. **101**, No. 2, p. 567—575 (1987).

G. Brauer and K. Popp
Neutron Embrittlement of Reactor Pressure Vessel Steels: A Challenge to Positron Annihilation and Other Methods
(61.80 and 78.70; 62.20; S1.1)
(a) Vol. **102**, No. 1, p. 79—90 (1987).

A. Seeger, J. Major, and F. Banhart
The Study of Positron Diffusion in Solids by Positron Spin Relaxation (e^+SR) Experiments
(66.30 and 71.60; 75.50; S1.1)
(a) Vol. **102**, No. 1, p. 91—106 (1987).

R. Würschum, M. Scheytt, and H.-E. Schaefer
Nanocrystalline Metals and Semiconductors Studied by Positron Lifetime Spectroscopy
(61.55 and 78.70; 61.70; S1.1; S1.2; S1.3; S5.11)
(a) Vol. **102**, No. 1, p. 119—126 (1987).

K. Ghazi-Wakili, U. Zimmermann, J. Brunner, P. Tipping, W. B. Waeber, and F. Heinrich
Positron Annihilation Studies on Neutron Irradiated Pressure Vessel Steels
(61.80 and 78.70; 62.20; 64.75; S1.1)
(a) Vol. **102**, No. 1, p. 153—163 (1987).

E. Kolawa, T. Tyliszczak, B. D. Sawicka, J. A. Sawicki, and M. Drwiega
Studies of Local Structure and Thermal Annealing Processes in Implanted Iron–Gadolinium Alloys by Conversion Electron Mössbauer Spectroscopy
(61.40; 61.70; 76.80; S1.1; S1.4)
(a) Vol. **102**, No. 1, p. 181—192 (1987).

S. P. Chenakin, V. T. Cherepin, A. L. Pivovarov, and M. A. Vasilev
Temperature Dependence of Secondary Ion Emission from Iron-Based Crystalline and Amorphous Alloys
(61.55 and 79.20; S1.1)
(a) Vol. **102**, No. 1, p. 231—240 (1987).

J. Baszyński
Magnetic Surface Anisotropy of the Fe–Cu Interface in Superlattices
(75.50 and 75.70; 76.50; S1.1; S1.2)
(b) Vol. **142**, No. 2, p. K165—K168 (1987).

D. Schödlbauer, G. Kögel, P. Sperr, and W. Triftshäuser
Lifetime Measurements with a Pulsed Slow Positron Beam
(78.70; S1.1; S1.2)
(a) Vol. **102**, No. 2, p. 549—554 (1987).

N. S. Saleh and K. A. Al-Saleh
Measurement of K-Shell X-Ray Cross Sections of Selected Elements from Ti to Zn for Incident Protons
(61.80; 78.70; S1.1; S1.2)
(a) Vol. **102**, No. 2, p. 619—623 (1987).

Y. B. XIA and Z. G. WANG
Evolution of Dislocation Structures during Low Cycle Fatigue in a Cr–Mn–N Stainless Steel
(61.70 and 62.20; S1.1)
(a) Vol. **103**, No. 2, p. 389—395 (1987).

A. I. C. PERSIANO and R. D. RAWLINGS
A Mössbauer Investigation of Equiatomic FeCo with Vanadium and Niobium Additions
(75.50 and 76.80; S1.1; S1.2)
(a) Vol. **103**, No. 2, p. 547—556 (1987).

R. BERTHE, A. BIRKNER, and U. HARTMANN
Current-Induced Changes of the Magnetic Domain Structure of Iron Whiskers
(75.60; S1.1)
(a) Vol. **103**, No. 2, p. 557—564 (1987).

H. BOSSE and H. GÄRTNER
A Systematic Ferromagnetic Resonance Study of Fe-, Ni-, and Cr-Ion Plated Amorphous Ribbons
(75.50 and 76.50; S1.1; S1.2)
(a) Vol. **103**, No. 2, p. 565—570 (1987).

C. GIOVANELLA, M. IANNUZZI, F. GAUZZI, G. PETROCCO, G. PRINCIPI, and P. Q. ZHANG
Production of Structurally and Magnetically Disordered $Fe_{20}Cu_{80}$ Metastable Alloys
(76.80; S1.1; S1.2)
(a) Vol. **103**, No. 2, p. K125—K129 (1987).

U. KREY, H. OSTERMEIER, and J. ZWECK
Electronic Properties of Glassy Metals
(71.20; 72.15; S1.1; S1.2)
(b) Vol. **144**, No. 1, p. 203—212 (1987).

W.-U. KELLNER, M. FÄHNLE, H. KRONMÜLLER, and S. N. KAUL
The Magnetic Phase Transition in Amorphous and Disordered Crystalline Ferromagnets
(75.30; 75.40; 75.50; S1.1; S1.2)
(b) Vol. **144**, No. 1, p. 397—409 (1987).

B. ŠESTÁK, Z. VICHERKOVÁ, V. NOVÁK, S. LIBOVICKÝ, and J. BRÁDLER
Cyclic Plastic Deformation of Fe–13.6 wt% Cr Single Crystals
(62.20; 61.70; S1.1)
(a) Vol. **104**, No. 1, p. 79—93 (1987).

D. BRUNNER and J. DIEHL
The Use of Stress-Relaxation Measurements for Investigations on the Flow Stress of α-Iron
(62.20; S1.1)
(a) Vol. **104**, No. 1, p. 145—155 (1987).

S1.2. Other 3d metals

A. Ostrasz and M. Szuskiewicz
3d-Like Electron Localization and Charge Transfer in Hydrogenated V–Fe Alloys
(71.25; 61.55; 76.80; S1.1; S1.2)
(b) Vol. **139**, No. 1, p. 205—211 (1987).

H. Solbrig
Recursive Scattered-Wave Approach to the Electronic Density of States of Topologically Disordered Systems
(71.20; S1.2)
(b) Vol. **139**, No. 1, p. 223—231 (1987).

R. Wang and D. Liu
Foil Thickness Determination by CBED in the Kinematical Approximation
(61.14; S1.2; S4; S10.1)
(a) Vol. **99**, No. 1, p. 47—55 (1987).

A. I. Ustinov, A. Yu. Gaevskii, A. D. Rood, V. S. Skorodzievskii, and K. V. Chuistov
The Influence of the Electron Subsystem and Kinetic Factors on the Formation of One-Dimensionally Disordered Structures with Long-Range Correlation
(61.50; S1.2)
(a) Vol. **99**, No. 1, p. 81—90 (1987).

He Yizhen and Li Xiao-Guang
A New Type of Internal Friction Peak of Metallic Glasses near T_g
(62.40; S1.2; S1.61)
(a) Vol. **99**, No. 1, p. 115—120 (1987).

I. Retat
Plastic Deformation of Fe–Ni and $Fe_{65}(Ni_{1-x}Mn_x)_{35}$ between 30 and 460 K. Magnetically Induced Invar Hardening
(62.20; 75.80; S1.1; S1.2)
(a) Vol. **99**, No. 1, p. 121—130 (1987).

J. M. Barandiarán, J. C. Gómez-Sal, J. Rodríguez Fernández, R. J. López Sánchez, and O. V. Nielsen
Temperature Dependence of the Electrical Resistivity in $[Co_{1-x}(Fe_{0.5}Ni_{0.5})_x]_{75}Si_{15}B_{10}$ Metallic Glasses
(72.15; S1.1; S1.2)
(a) Vol. **99**, No. 1, p. 243—249 (1987).

I. Okamoto, T. Miyazaki, and M. Takahashi
Change of Magnetic Properties with Heat Treatments in Rapidly Quenched $Cu_{99.2}Fe_{0.8}$ Alloys
(75.20 and 75.50; S1.1; S1.2)
(a) Vol. **99**, No. 1, p. 267—274 (1987).

W. Wegener and C. Bostoen
Investigation of Hydrogen Diffusion in CuH by Means of Quasielastic Neutron Scattering
(66.30; S1.2)
(a) Vol. **99**, No. 1, p. K19—K21 (1987).

P. GODOWSKI and J. RUDNY
Effect of Sulphur on the Surface Segregation in $Co_{50}Ni_{50}$ Alloy
(68.20; 79.20; S1.2)
(a) Vol. **99**, No. 2, p. 527—536 (1987).

S. DE STEFANO, A. DI CHIARA, G. PELUSO, and U. SCOTTI DI UCCIO
Aspects of Magnetic Penetration in Vanadium Based Josephson Junctions
(74.30 and 74.50; S1.2)
(a) Vol. **99**, No. 2, p. K109—K110 (1987).

V. S. DEMIDENKO, A. SZÁSZ, and M. A. AYSAWI
On the Model Calculation of the Excitonic-Like States and Their Possible Role in Autocatalytic Processes. I. Application to the Formation of Autocatalytic Ni–P Amorphous Coatings
(68.45 and 68.55; 71.35; S1.2; S1.62)
(b) Vol. **140**, No. 1, p. 121—126 (1987).

S. SZUSZKIEWICZ
A Study of the Electronic Structure of $Fe_{1-x}Co_x$ Alloys by a Polarized Positron Annihilation
(71.25; 78.70; S1.1; S1.2)
(b) Vol. **140**, No. 1, p. 141—147 (1987).

L. M. GRATTON, A. MIOTELLO, and C. TOSELLO
Thermal Energy Balances in Ion Implanted Copper Bars
(61.80; 65.50; S1.2)
(a) Vol. **100**, No. 1, p. 53—57 (1987).

G. VANDERSCHAEVE
Cross-Slip of Partial Dislocations via the Stair Rod Mode. The Zigzag Propagation of Deformation Microtwins in Ordering Alloys
(61.70; S1.2)
(a) Vol. **100**, No. 1, p. 59—68 (1987).

W. TH. M. BUTERS, A. VAN VEEN, and A. VAN DEN BEUKEL
Some Results on Helium Trapping in Undeformed and Cold Worked Single Crystalline Copper. A Comparison with Molybdenum and Nickel
(61.70; 68.45; 79.20; S1.2)
(a) Vol. **100**, No. 1, p. 87—94 (1987).

L. KRAUS, I. TOMÁŠ, E. KRATOCHVÍLOVÁ, B. SPRINGMANN, and E. MÜLLER
Magnetic Anisotropy Caused by Oriented Surface Roughness of Amorphous Ribbons
(75.30; 75.50; S1.1; S1.2)
(a) Vol. **100**, No. 1, p. 289—299 (1987).

W. ANDRÄ and H. DANAN
Conditions for Perpendicular Magnetic Recording in Terms of the Wall-Displacement Model
(75.70; S1.2)
(a) Vol. **100**, No. 1, p. 307—316 (1987).

S. SAITO and T. MAEDA
Temperature Variation of the Work Function of Polycrystalline Cobalt in the Temperature Range $100\ ^\circ C \leqq T \leqq 485\ ^\circ C$
(73.30; S1.2)
(a) Vol. **100**, No. 1, p. K21—K24 (1987).

J. Kierul, J. W. Stasiak, and M. Kluczynski
Correlation and Influence of Phonons in Calculations of the Density of States in Thin Films
(71.20; 71.38; S1.2)
(b) Vol. **141**, No. 1, p. 119—128 (1987).

B. Lippold, H. Müller, and S. Mahmoud
Quadrupole Scattering in $PrNi_5$
(72.15; S1.2; S1.4)
(b) Vol. **141**, No. 1, p. 247—251 (1987).

T. Soma, T. Hosaka, and H.-Matsuo Kagaya
Temperature Dependence of the Elastic Constants for Cu
(62.20; S1.2)
(b) Vol. **141**, No. 1, p. K21—K24 (1987).

R. Krishnan and M. Dancygier
High Field Susceptibility Peak in Some NiFeSiB Amorphous Alloys
(75.60; S1.1; S1.2)
(b) Vol. **141**, No. 1, p. K71—K74 (1987).

M. Szuszkiewicz, S. Szuszkiewicz, and J. Chojcan
Back-Scattering of Positrons on Magnetized Samples
(78.70; S1.1; S1.2)
(b) Vol. **141**, No. 1, p. K79—K81 (1987).

T. Balcerzak and J. P. Jakubovics
Studies of Fe–Ni and Fe–Ni–Gd Thin Films by Lorentz Microscopy
(75.70; 68.55; S1.1; S1.2)
(a) Vol. **101**, No. 1, p. 217—225 (1987).

Z. Frait, D. Fraitová, N. Zárubová, and P. Duhaj
Magnetic Surface Anisotropy and Ferromagnetic Resonance in a Stress Annealed FeNiB Amorphous Alloy
(76.50; 75.30; S1.1; S1.2)
(a) Vol. **101**, No. 1, p. 239—245 (1987).

T. M. Zhantikin, A. K. Zhetbaev, M. F. Vereshchak, and Z. T. Serikbaeva
Mössbauer Investigations of Radiation Damage in Ni_3Fe
(61.80; 76.80; S1.1; S1.2)
(a) Vol. **101**, No. 2, p. 361—370 (1987).

A. Zeghib, J. Le Bas, and C. Vautier
Electrical and Magnetic Properties of Amorphous Ni_xAg_{1-x} Thin Films
(73.60 and 75.70; S1.2; S1.3)
(a) Vol. **101**, No. 2, p. 487—493 (1987).

D. I. Sementsov and V. V. Sidorenkov
Ferromagnetic Resonance in Films with Angular Dispersion of Magnetic Anisotropy
(75.70; 76.50; S1.1; S1.2)
(a) Vol. **101**, No. 2, p. 555—562 (1987).

W. Lühr-Tanck, A. Sager, M. Ederhof, and H. Bosse
A Positron Annihilation Study of Precipitation Effects in CuIn and CuSb Alloys
(64.75 and 78.70; 61.70; S1.2)
(a) Vol. **102**, No. 1, p. 133—138 (1987).

Zhu Guang-qu, Lu Shan-yi, and Wang Shao-jie
Investigation of Microstructure and Coercivity Mechanisms in the Permanent Magnet Alloy Mn–Al–C by Positron Annihilation and Transmission Electron Microscopy
(75.60 and 78.70; S1.2)
(a) Vol. **102**, No. 1, p. 165—169 (1987).

Moheb H. Youssef and M. Roushdy
A Modified Method for Measuring Ultrasonic Attenuation in Highly Attenuating Materials
(62.80; S1.2)
(a) Vol. **102**, No. 1, p. 287—292 (1987).

P. N. Ram and P. D. Semalty
Calculated Elastic Constants of Dilute Alloys
(62.20; S1.2; S1.3)
(b) Vol. **142**, No. 2, p. 387—399 (1987).

C. Chaîb and J.-G. Gasser
The Temperature Dependence of the Resistivity of Liquid Copper
(72.15; 61.25; S1.2)
(b) Vol. **142**, No. 2, p. 531—536 (1987).

J. Baszyński
Magnetic Surface Anisotropy of the Fe–Cu Interface in Superlattices
(75.50 and 75.70; 76.50; S1.1; S1.2)
(b) Vol. **142**, No. 2, p. K165—K168 (1987).

A. Ostrasz and M. Szuszkiewicz
Comment on the Magnetic Moments Formation in Hydrogenated V–Fe Alloys
(76.80; S1.2)
(b) Vol. **142**, No. 2, p. K173—K176 (1987).

D. Schödlbauer, G. Kögel, P. Sperr, and W. Triftshäuser
Lifetime Measurements with a Pulsed Slow Positron Beam
(78.70; S1.1; S1.2)
(a) Vol. **102**, No. 2, p. 549—554 (1987).

G. Van Tendeloo, D. Van Dyck, S. Kuypers, H. W. Zandbergen, and S. Amelinckx
Electron Diffraction Effects in Mixed Layer Compounds. II. Applications
(61.14; S1.2; S8.16)
(a) Vol. **102**, No. 2, p. 597—608 (1987).

N. S. Saleh and K. A. Al-Saleh
Measurement of K-Shell X-Ray Cross Sections of Selected Elements from Ti to Zn for Incident Protons
(61.80; 78.70; S1.1; S1.2)
(a) Vol. **102**, No. 2, p. 619—623 (1987).

W. SCHILLER
On the Extension of the Parallel-Resistor Model for High-Resistivity Conductors
(72.15; S1.2)
(b) Vol. **143**, No. 1, p. K47—K52 (1987).

S. H. SINHA and A. R. CHETAL
Correlation between Chemical Shift, Plasmon Energy, and Partial Charge
(71.45; S1.2)
(b) Vol. **143**, No. 2, p. 595—600 (1987).

A. YU. RUMYANTSEV, N. L. MITROFANOV, A. S. IVANOV, N. A. CHERNOPLEKOV, L. WEISS, H. ESCHRIG, and R. RICHTER
Anomalies in the Phonon Spectra and the Fermi Surface of the Random Substitutional Alloy $Cu_{0.9}Ni_{0.1}$
(63.50 and 71.25; S1.2)
(b) Vol. **143**, No. 2, p. 625—636 (1987).

Z. BADIRKHAN and J. KHALIFEH
Charge Transfer in Nickel–Hydrogen System
(71.55; S1.2)
(b) Vol. **143**, No. 2, p. 637—645 (1987).

I. KANAZAWA, H. MURAKAMI, and M. DOYAMA
The Migration Stage of Divacancies in Quenched Pure Aluminium, Dilute Aluminium Alloys, and Pure Copper Studied by Positron Annihilation
(61.70; 78.70; S1.2; S4)
(a) Vol. **103**, No. 2, p. 403—411 (1987).

A. I. C. PERSIANO and R. D. RAWLINGS
A Mössbauer Investigation of Equiatomic FeCo with Vanadium and Niobium Additions
(75.50 and 76.80; S1.1; S1.2)
(a) Vol. **103**, No. 2, p. 547—556 (1987).

H. BOSSE and H. GÄRTNER
A Systematic Ferromagnetic Resonance Study of Fe-, Ni-, and Cr-Ion Plated Amorphous Ribbons
(75.50 and 76.50; S1.1; S1.2)
(a) Vol. **103**, No. 2, p. 565—570 (1987).

C. GIOVANELLA, M. IANNUZZI, F. GAUZZI, G. PETROCCO, G. PRINCIPI, and P. Q. ZHANG
Production of Structurally and Magnetically Disordered $Fe_{20}Cu_{80}$ Metastable Alloys
(76.80; S1.1; S1.2)
(a) Vol. **103**, No. 2, p. K125—K129 (1987).

D. N. SEIDMAN, R. S. AVERBACK, and R. BENEDEK
Displacement Cascades. Dynamics and Atomic Structure
(61.80; 61.70; S1; S1.2)
(b) Vol. **144**, No. 1, p. 85—104 (1987).

H.-G. FRITSCHE
A Comparison of Electronic Properties of Clusters and Bulk of Copper and Palladium. Bandwidth, Density of States, Deformation Density
(71.20; 71.25; S1; S1.2)
(b) Vol. **144**, No. 1, p. 195—202 (1987).

M. Wilkens, T. Ungár, and H. Mughrabi
X-Ray Rocking-Curve Broadening of Tensile-Deformed [001]-Orientated Copper Single Crystals
(61.10 and 61.70; S1.2)
(a) Vol. **104**, No. 1, p. 157—170 (1987).

A. Hampel and H. Neuhäuser
Investigation of Slip Line Growth in F.C.C. Cu Alloys with High Resolution in Time
(61.70; 62.20; S1.2)
(a) Vol. **104**, No. 1, p. 171—181 (1987).

H. M. Ledbetter and M. W. Austin
Deformed-Polycrystalline-Copper Elastic Constants
(62.20; 61.70; S1.2)
(a) Vol. **104**, No. 1, p. 203—212 (1987).

H. Mehrer
Implantation of Radioisotopes at ISOLDE. A Novel Source Production Technique for Diffusion Studies in Solids
(66.30; 61.70; S1.2; S1.3; S5.11)
(a) Vol. **104**, No. 1, p. 247—261 (1987).

B. C. Larson and F. W. Young, Jr.
X-Ray Diffuse Scattering Study of Irradiation Induced Dislocation Loops in Copper
(61.10 and 61.80; S1.2)
(a) Vol. **104**, No. 1, p. 273—286 (1987).

M.-P. Macht, A. Müller, V. Naundorf, and H. Wollenberger
Determination of Temperature-Dependent Sink Concentrations in Ion-Irradiated Nickel and Copper by Diffusion Measurements
(61.70; 61.80; 66.30; S1.2)
(a) Vol. **104**, No. 1, p. 287—295 (1987).

P. Linden, P. Lamparter, and S. Steeb
Crystallization and Relaxation of Amorphous Mg–Zn, Cu–Ti, and Ni–P Alloys under High Hydrostatic Pressure
(61.40 and 62.50; 72.15; S1.2; S3)
(a) Vol. **104**, No. 1, p. 357—368 (1987).

N. Willmann, W. Mader, E. Wachtel, and B. Predel
Decomposition and Crystallization Behaviour of the Amorphous $Ni_{39}Pd_{43}P_{18}$ and $Ni_{37}Pd_{45}P_{18}$ Alloys
(61.40; 64.80; 75.50; S1.2)
(a) Vol. **104**, No. 1, p. 369—379 (1987).

H. Indrawirawan, O. Buck, and O. N. Carlson
Substitutional–Interstitial Solute Interactions in Niobium–Vanadium–Oxygen Alloys
(62.40; 61.70; 66.30; S1; S1.2)
(a) Vol. **104**, No. 1, p. 443—451 (1987).

H. Mizubayashi, Y. Katoh, and S. Okuda
Hydrogen Internal Friction Peak and the Effect of Quenching in Amorphous CuTi
(61.40 and 62.40; S1.2)
(a) Vol. **104**, No. 1, p. 469—485 (1987).

S1.3. Noble metals

H. P. BONDE and S. B. SHRIVASTAVA
Electron Momentum Density Distributions in Rhodium
(71.25; 78.70; S1; S1.3)
(b) Vol. **139**, No. 1, p. 155—160 (1987).

J. BANHART, H. EBERT, and J. VOITLÄNDER
Application of the Coherent Potential Approximation to Substitutional Ternary Alloys
(71.20; S1; S1.3)
(b) Vol. **139**, No. 1, p. K19—K23 (1987).

M. PROCOP, H. RAIDT, and B. SANDOW
Auger Electron and X-Ray Spectroscopic Analysis of Au–Ge Contacts to GaAs
(68.48; 73.40; 79.20; S1.3; S7.12)
(a) Vol. **99**, No. 2, p. 573—580 (1987).

R. V. GOPALA RAO and M. BHATTACHARYYA
Partial Structures of Compound-Forming Liquid Gold–Caesium Alloy. A Model Potential Approach
(61.25 and 61.55; S1.3; S2)
(b) Vol. **140**, No. 1, p. 51—61 (1987).

A. A. BEDNYAKOV, G. A. IFEROV, A. G. KADMENSKII, N. M. NAUMOVA, and YU. N. ZHUKOVA
The Role of Single Collisions in the Process of Scattering and Stopping of 200 to 400 keV Protons in Thin Gold Foils
(61.80; S1.3)
(b) Vol. **140**, No. 1, p. 63—71 (1987).

A. J. ABU EL-HAIJA and A. H. MADJID
DC Electrical Conductivity of Ultrathin Ag–SiO Superlattices
(73.40; S1.3; S10.1)
(b) Vol. **140**, No. 1, p. 113—120 (1987).

T. BHATTACHARYA and A. K. GHOSH
Pair-Correlation Functions of Liquid Noble Metals through Born-Green-Yvonne Theory
(61.25; 65.50; S1.2; S1.3)
(b) Vol. **140**, No. 2, p. 331—334 (1987).

B. K. SHARMA, HANUMAN SINGH, S. PERKKIÖ, T. PAAKKARI, and K. MANSIKKA
Electron Momentum Distribution in Silver and the Renormalized Free Atom Model
(71.25; 78.70; S1.3)
(b) Vol. **141**, No. 1, p. 177—184 (1987).

A. ZEGHIB, J. LE BAS, and C. VAUTIER
Electrical and Magnetic Properties of Amorphous Ni_xAg_{1-x} Thin Films
(73.60 and 75.70; S1.2; S1.3)
(a) Vol. **101**, No. 2, p. 487—493 (1987).

C. Malinowska-Adamska
Self-Consistent Calculations of the Melting Criterion Based on Lattice Instability
(63.20 and 64.70; S1.2; S1.3; S2; S4)
(b) Vol. **143**, No. 1, p. 63—71 (1987).

T. Soma, S. Suzuki, and H.-Matsuo Kagayá
Pressure Effect on the Melting Point of Noble Metals
(63.20; 65.70; S1.3)
(b) Vol. **143**, No. 1, p. 99—103 (1987).

L. Wendler and R. Haupt
An Improved Virtual Mode Theory of ATR Experiments on Surface Polaritons. Application to Long-Range Surface Plasmon-Polaritons in Asymmetric Layer Structures
(68.30 and 78.65; 71.36; 71.45; S1.3)
(b) Vol. **143**, No. 1, p. 131—148 (1987).

C. Schober and V. N. Antonov
The Gyromagnetic Factor of Conduction Electrons in Noble and Transition Metals
(71.25; S1; S1.3)
(b) Vol. **143**, No. 1, p. K31—K35 (1987).

R. V. G. Rao and M. Bhattacharyya
Caesium-Gold as an Ionic Alloy and Its Structural Properties
(61.55; S1.3; S2)
(a) Vol. **103**, No. 1, p. 79—84 (1987).

K. Yamamoto and S. Maruyama
A New Long Period Structure in Gold–Indium Films between 20 and 25at% In
(68.55; S1.3; S4)
(a) Vol. **103**, No. 1, p. 153—160 (1987).

Xueming Ma and Yizhen He
Viscous Flow Behaviour of Metallic Glass $Pd_{77.5}Ni_6Si_{16.5}$ near T_g
(62.20; S1.3)
(a) Vol. **103**, No. 1, p. K13—K16 (1987).

G. Vogl
On the Motion of Solute Atoms in Cages and Loops in Metal Lattices Studied with Nuclear Methods
(61.70; 66.30; 76.80; S1; S1.3; S4)
(b) Vol. **144**, No. 1, p. 259—269 (1987).

G. Neumann
A Model for the Calculation of Monovacancy and Divancancy Contributions to the Impurity Diffusion in Noble Metals
(61.70 and 66.30; S1.2; S1.3)
(b) Vol. **144**, No. 1, p. 329—341 (1987).

H. Mehrer
Implantation of Radioisotopes at ISOLDE. A Novel Source Production Technique for Diffusion Studies in Solids
(66.30; 61.70; S1.2; S1.3; S5.11)
(a) Vol. **104**, No. 1, p. 247—261 (1987).

H.-E. Schaefer and F. Banhart
Thermal Equilibrium Vacancies in Platinum Studied by Positron Annihilation
(61.70; 71.60; 78.70; S1.3)
(a) Vol. **104**, No. 1, p. 263—272 (1987).

A. Bartels, D. Bartusel, and K. Lücke
On the Nature of Frenkel Defects in Concentrated AuAg Alloys
(61.70; 72.15; S1.3)
(a) Vol. **104**, No. 1, p. 315—328 (1987).

T. Shiozawa and T. Kobayashi
High Resolution Electron Microscopic Study of Photolytic Silver in Silver Bromide
(61.80; 61.16; S1.3; S9)
(a) Vol. **104**, No. 2, p. 649—659 (1987).

S1.4. Lanthanides

S. L. Kakani and U. N. Upadhyaya
Coexistence of Superconductivity and Ferromagnetism in Rare Earth Ternary Systems **(Review Article)**
(74.10 and 75.30; S1.4)
(a) Vol. **99**, No. 1, p. 15—36 (1987).

A. Rojek and C. Sułkowski
Anomalous Magnetic and Electrical Properties of the $EuRh_{1.5}Sn_{4.2}$ Compound
(72.15 and 75.60; S1.4)
(b) Vol. **139**, No. 1, p. K63—K67 (1987).

K.-H. Müller, A. Handstein, D. Eckert, and J. Schneider
The Dip in Magnetization Curves of Sintered Nd–Fe–B Permanent Magnets
(75.60; S1.1; S1.4)
(a) Vol. **99**, No. 1, p. K61—K64 (1987).

A. Urbaniak-Kucharczyk
Spin Autocorrelation Time in Thin Iron, Nickel, and Gadolinium Films
(78.70; S1.1; S1.2; S1.4)
(b) Vol. **139**, No. 2, p. K141—K143 (1987).

M. M. Amado, J. D. Montenegro, J. M. Moreira, M. E. Braga, and J. B. Sousa
Study of the Magnetic Transitions in $CeFe_2$ with High Resolution Transport Property Measurements
(75.40; 72.15; 75.50; S1.1; S1.4)
(a) Vol. **99**, No. 2, p. 625—630 (1987).

T. D. Radzhabov, A. M. Nazarov, and U. U. Pulatov
Surface Properties of Ion Alloyed Thin Dysprosium Films
(68.45; S1.4)
(a) Vol. **99**, No. 2, p. K83—K86 (1987).

K. Tabata and H. Kido
Preparation and Magnetic Properties of $La_{1-x}Th_xCoO_3$ Compounds
(75.50; S1.4)
(a) Vol. **99**, No. 2, p. K121—K123 (1987).

Yu. M. Seidov and A. M. Suleimanov
To the Theory of Two-Magnon Absorption of Light by Surface Collective Excitations in the Singlet–Triplet Ferromagnets
(71.70; 75.30; 78.20; S1.4)
(b) Vol. **140**, No. 1, p. 149—154 (1987).

M. Budzyński, O. I. Kochetov, M. Subotowicz, H. Niezgoda, H. Spustek, and W. Tańska-Krupa
Influence of Hydrogenation on the Hyperfine Interactions in $(Y_{0.97}Hf_{0.03})Fe_2$
(75.20; 75.25; 75.50; 76.80; S1.1; S1.4)
(b) Vol. **140**, No. 2, p. 589—594 (1987).

B. Lippold, H. Müller, and S. Mahmoud
Quadrupole Scattering in $PrNi_5$
(72.15; S1.2; S1.4)
(b) Vol. **141**, No. 1, p. 247—251 (1987).

C. Weber, M. Wagner, and E. Sigmund
Flexibility Options in Valency-Mixing Jahn-Teller Systems
(71.70 and 76.30; S1.4)
(b) Vol. **141**, No. 2, p. 529—539 (1987).

G. Fischer and A. Herr
Representation of Energetical and Low-Field Magnetic Properties of $J = 5/2$ Rare Earth Ion States in Tetragonal Crystal Fields
(75.10; S1.4)
(b) Vol. **141**, No. 2, p. 589—598 (1987).

D. Eckert, D. Hinz, A. Handstein, and J. Schneider
Thermal Ageing and Coercivity of Sintered Nd–Fe–B Magnets
(75.50; 75.40; 75.60; S1.1; S1.4)
(a) Vol. **101**, No. 2, p. 563—566 (1987).

E. Kolawa, T. Tyliszczak, B. D. Sawicka, J. A. Sawicki, and M. Drwiega
Studies of Local Structure and Thermal Annealing Processes in Implanted Iron–Gadolinium Alloys by Conversion Electron Mössbauer Spectroscopy
(61.40; 61.70; 76.80; S1.1; S1.4)
(a) Vol. **102**, No. 1, p. 181—192 (1987).

R. Scholl, L. Jahn, and R. Schumann
The Thermal Remagnetization in 2:17 Sm–Co Permanent Magnets
(75.60; S1.4)
(a) Vol. **102**, No. 1, p. K37—K41 (1987).

C. F. van Bruggen
Monovalent-Copper (I; $3d^{10}$, $S = 0$) Induced Oxygen-2p Valence Band Holes with Concomitant Attractive Interaction inside Low-Dimensional Oxygen Networks
(71.25 and 74.70; 75.20; S1.4; S10.15)
(b) Vol. **142**, No. 2, p. 537—550 (1987).

M. Kurisu, H. Kadomatsu, and H. Fujiwara
Magnetic Field Dependence of Two Kinds of Structural Transition Temperatures of $CeAg_{1-x}In_x$
(64.60 and 75.30; 72.15; S1.3; S1.4)
(b) Vol. **142**, No. 2, p. 581—585 (1987).

Z. Kąkol, H. Figiel, and Cz. Kapusta
The Role of Gadolinium in Magnetocrystalline Anisotropy of $(Y_{1-x}Gd_x)_2Co_{17}$ Pseudobinaries
(75.50; S1.2; S1.4)
(b) Vol. **143**, No. 1, p. 255—260 (1987).

J. Völkl, H. Wipf, B. J. Beaudry, and K. A. Gschneidner, Jr.
Diffusion of H and D in Lutetium
(66.30; S1.4)
(b) Vol. **144**, No. 1, p. 315—327 (1987).

R. N. Joarder and T. Das
The Thermodynamic Properties of Liquid Transition and Rare-Earth Metals Based on CHS Model
(65.50; 65.20; S1.2; S1.4)
(b) Vol. **144**, No. 2, p. 557—564 (1987).

Hanmin Jin, Guanghua Guo, Ensheng Chen, Yong Zhu, Riuwen Zhao, Shinwen Li, and Fuming Yang
Magnetization Process in Nd–Fe–B Permanent Magnets near Remanence at Low Temperatures
(75.50; 75.60; S1.1; S1.4)
(a) Vol. **104**, No. 2, p. 799—804 (1987).

S1.5. Actinides

A. Ghorai and S. K. Sen
Calculations on Relaxations Due to a Vacancy, Binding Energy of a Vacancy–Vacancy Pair, and Formation Energy of a Vacancy in Thorium
(61.70; 63.20; S1.5)
(b) Vol. **139**, No. 1, p. 77—83 (1987).

N. E. Alekseevski, A. V. Mitin, E. P. Khlybov, A. Gilewski, and B. Greń
The Influence of the "Quality" of the Specimens on the Galvanomagnetic Properties of UBe_{13}
(72.15; S1.5)
(a) Vol. **100**, No. 2, p. K165—K168 (1987).

B. N. Onwuagba
(Sd)–f Hybridization in the Spectrum of Thorium
(63.20; S1.5)
(b) Vol. **141**, No. 1, p. 105—110 (1987).

S1.61. Carbides, silicides, germanides

A. L. Ivanovskii, D. L. Novikov, and V. A. Gubanov
Electronic Properties of Carbides, Nitrides, and Oxides of Subgroup VIa Transition Metals. Theory and Experiment **(Review Article)**
(71.25; 73.20; 78.70; 79.60; S1.1; S1.62; S10)
(b) Vol. **141**, No. 1, p. 9—33 (1987).

D. L. Novikov, A. L. Ivanovskii, and V. A. Gubanov
The Influence of Carbon Vacancies and Local Atomic Displacements on the Electronic Structure of Niobium Carbides
(71.25 and 71.55; S1.61)
(b) Vol. **139**, No. 1, p. 257—265 (1987).

R. B. Prabhu
Mix-Valent Europium Based Systems in Interconfiguration Fluctuation Model
(75.10; S1.61; S1.62)
(b) Vol. **139**, No. 1, p. K69—K73 (1987).

He Yizhen and Li Xiao-Guang
A New Type of Internal Friction Peak of Metallic Glasses near T_g
(62.40; S1.2; S1.61)
(a) Vol. **99**, No. 1, p. 115—120 (1987).

R. Portius, D. Dietrich, A. Hässner, Th. Raschke, E. Wieser, and W. Wolke
Formation of Stable $MoSi_2$/Poly-Si Films by Rapid Thermal Annealing
(68.48 and 73.60; S1.61; S5.11)
(a) Vol. **100**, No. 1, p. 199—206 (1987).

A. Singh and W. S. Khokle
Sheet Resistance and Grain Size Study of Nickel Disilicide
(68.55 and 73.60; S1.61)
(a) Vol. **100**, No. 1, p. K25—K29 (1987).

A. I. Gusev and A. A. Rempel
Calculating the Energy Parameters for the CV and OPF Methods
(61.50; 64.60; S1.61)
(b) Vol. **140**, No. 2, p. 335—346 (1987).

Chan-Syun Yang and Juh-Tzeng Lue
Palladium Silicide Formation on Amorphous Silicon by Thermal Annealing
(68.55; 68.48; S1.61)
(a) Vol. **101**, No. 1, p. 151—157 (1987).

H. Cebulla, E. Erben, T. Gessner, and E. Vetter
Induced Tantalum Silicide Formation by Ar^+ Ion Implantation
(61.70 and 68.55; S1.61)
(a) Vol. **101**, No. 2, p. K125—K127 (1987).

S1.62. Nitrides, phosphides, arsenides

A. L. Ivanovskii, D. L. Novikov, and V. A. Gubanov
Electronic Properties of Carbides, Nitrides, and Oxides of Subgroup VIa Transition Metals. Theory and Experiment **(Review Article)**
(71.25; 73.20; 78.70; 79.60; S1.61; S1.62; S10)
(b) Vol. **141**, No. 1, p. 9—33 (1987).

R. B. Prabhu
Mix-Valent Europium Based Systems in Interconfiguration Fluctuation Model
(75.10; S1.61; S1.62)
(b) Vol. **139**, No. 1, p. K69—K73 (1987).

V. S. Demidenko, A. Szász, and M. A. Aysawi
On the Model Calculation of the Excitonic-Like States and Their Possible Role in Autocatalytic Processes. I. Application to the Formation of Autocatalytic Ni–P Amorphous Coatings
(68.45 and 68.55; 71.35; S1.2; S1.62)
(b) Vol. **140**, No. 1, p. 121—126 (1987).

I. F. Gribanov and E. A. Zavadskii
Magnetovolume Effects in MnAs. Role of Space Spin Correlations
(75.80; S1.62)
(b) Vol. **142**, No. 2, p. 559—568 (1987).

H. Matsui, S. Suzuki, and M. Tamaki
Characteristic Temperatures in the Electrical Resistivity and Magnetization of β-U_2N_3
(72.15 and 75.50; S1.62)
(a) Vol. **103**, No. 1, p. 199—203 (1987).

P. Čapková, L. Smrčok, V. Šíma, V. Valvoda, B. Chalupa, and P. Ettmayer
Thermal Vibrations in Substoichiometric Vanadium Nitrides. Dynamical Deformation Effect
(63.20; 65.70; S1.62)
(b) Vol. **143**, No. 2, p. 471—476 (1987).

H. Matsui
Characteristic Temperatures of Uranium Monophosphide (UP) in Magnetic Susceptibility and Electrical Resistivity
(72.15 and 75.30; S1.62)
(a) Vol. **104**, No. 2, p. 825—831 (1987).

S1.63. Sulfides, selenides, tellurides

O. GOEDE, W. HEIMBRODT, V. WEINHOLD, E. SCHNÜRER, and H. G. EBERLE
Optical Study of Phase Transition from Tetrahedrally into Octahedrally Coordinated MnS
(78.65; 68.55; S1.63)
(b) Vol. **143**, No. 2, p. 511—518 (1987).

W. SIENICKI
Obtaining and Study of Tungsten Dichalcogenides Intercalated by Indium
(72.80; 61.70; S1.63)
(a) Vol. **103**, No. 2, p. K103—K105 (1987).

A. G. RUSTAMOV, G. I. GASHIMOV, and A. A. MUSTAFAEV
Investigation of the Electrophysical Properties of an NiS Single Crystal in the Magnetic-Transformation Region
(72.15 and 75.50; S1.63)
(a) Vol. **103**, No. 2, p. K131—K135 (1987).

H. L. PARK and W. B. WHITE
Raman Spectra of FeS and β-CoS
(78.30; S1.63)
(a) Vol. **144**, No. 1, p. K69—K71 (1987).

S2. Alkali metals and their alloys and compounds
(for alkaline halides, see S9.11)

T. SOMA, K. IWAFUNE, and H.-MATSUO KAGAYA
Contribution from the Lattice Vibrational Term and Pressure Dependence of the Phase Diagrams for Alkali Alloys
(63.20; 64.70; S2)
(b) Vol. **143**, No. 2, p. 477—485 (1987).

H. BROSS and R. STRYCZEK
Self-Consistent MAPW Calculation with a Warped Muffin-Tin Potential. II. The Electronic Structure of Li and Its Pressure Dependence
(71.20 and 71.25; 64.30; S2)
(b) Vol. **144**, No. 2, p. 675—692 (1987).

S3. Alkaline earth metals and their alloys and compounds
(for alkaline earth halides, see S9.12)

R. UBGADE and P. R. SARODE
Study of Strontium Compounds and Minerals by X-Ray Absorption Spectroscopy. I. K-Edge Shifts
(78.70; S3; S9.12; S10; S11.1)
(a) Vol. **99**, No. 1, p. 295—301 (1987).

F. POVOLO and R. E. BOLMARO
Average Elastic Constants and Tensor Invariants
(62.20; S1.2; S3; S11.1; S12)
(a) Vol. **99**, No. 2, p. 423—436 (1987).

E. BECKER and W. HEYROTH
Behaviour of Young's Modulus of AlZnMg Alloys during Homogeneous Decomposition
(62.20; S1.2; S3; S4)
(a) Vol. **100**, No. 2, p. 485—492 (1987).

H. C. GUPTA, G. S. REDDY, and B. B. TRIPATHI
Lattice Dynamics of B.C.C. Sr — A Pseudopotential Approach
(63.20; S3)
(b) Vol. **141**, No. 2, p. K97—K100 (1987).

J. C. UPADHYAYA and O. P. KULSHRESTHA
Lattice Dynamics of Ca in the Resonant Model Potential
(63.20; S3)
(b) Vol. **142**, No. 2, p. K109—K112 (1987).

P. LINDEN, P. LAMPARTER, and S. STEEB
Crystallization and Relaxation of Amorphous Mg–Zn, Cu–Ti, and Ni–P Alloys under High Hydrostatic Pressure
(61.40 and 62.50; 72.15; S1.2; S3)
(a) Vol. **104**, No. 1, p. 357—368 (1987).

S4. Other metals and their alloys

TRAN THOAI DUY BAO
Plasmon Excitation in the Interaction of Fast Electrons with Metal Surfaces
(71.45 and 73.20; 61.80; S4)
(b) Vol. **139**, No. 1, p. 145—154 (1987).

P. P. BODIUL, V. F. GARABAZHIU, and D. V. GITSU
The Impurity States in Bismuth
(71.55; 71.70; S4)
(b) Vol. **139**, No. 1, p. 245—256 (1987).

R. WANG and D. LIU
Foil Thickness Determination by CBED in the Kinematical Approximation
(61.14; S1.2; S4; S10.1)
(a) Vol. **99**, No. 1, p. 47—55 (1987).

H. LÖFFLER, G. WENDROCK, and K. PAWELZYK
The Influence of Small Mg Additions on the Stability of Guinier-Preston Zones in Al–4.5 at% Zn–x Mg Alloys
(61.55; 61.10; 64.70; S4)
(a) Vol. **99**, No. 1, p. 73—80 (1987).

XIAN-WU ZOU, ZHUN-ZHI JIN, and YA-JUN SHANG
Static Structure Factor of Non-Simple Liquid Metals Bi, Ga, Sb, and Sn
(61.25; S4)
(b) Vol. **139**, No. 2, p. 365—370 (1987).

J. ROSE and R. SCHUCHARDT
A Pseudopotential Band Structure Calculation of Bismuth, Antimony, and Their Alloys
(71.25; 71.30; 76.40; S4)
(b) Vol. **139**, No. 2, p. 499—518 (1987).

G. JUNGK and V. SCHULTZE
Optical Properties of Al Films within and beyond the Effective-Medium Approximation
(78.65; S4)
(b) Vol. **139**, No. 2, p. 627—633 (1987).

M. MEHAR KHAN and S. M. RAFIQUE
Electronic Transport Properties of Gallium: Harrison's Method
(72.15; S4)
(b) Vol. **139**, No. 2, p. K131—K135 (1987).

B. M. BOLOTOV, D. A. AKSENOV, A. I. POLYAKOV, and S. P. SENSHIN
NMR Line Shape Function Analysis in Solids
(76.60; S4; S9.11; S9.12)
(b) Vol. **139**, No. 2, p. K137—K140 (1987).

D. LIU, R. WANG, and Y. HUANG
Ultra-Thin Foil Thickness Determination from Large-Angle Convergent Beam Electron Diffraction Patterns
(61.14; S1.2; S4)
(a) Vol. **99**, No. 2, p. 339—344 (1987).

P. GONDI, A. SILI, and E. BONETTI
Relaxation Phenomena of Dislocations Contributing to Deformation of Al at Intermediate Temperatures
(61.70; 62.20; 62.40; S4)
(a) Vol. **99**, No. 2, p. 375—384 (1987).

J. MAYER, K. URBAN, and J. FIDLER
Phase Transitions between the Quasicrystalline, Crystalline, and Amorphous Phases in Al–16 at% V
(64.70; 61.40; S4)
(a) Vol. **99**, No. 2, p. 467—473 (1987).

T. P. EGGARTER and A. TROPER
A Simple Cluster Improvement to the Coherent Potential Approximation
(71.10; 71.25; S4; S12.1)
(b) Vol. **140**, No. 1, p. 127—133 (1987).

T. SATOW, O. UEMURA, K. MATSUMOTO, and S. OKAMURA
Electronic Properties of Liquid Sb–Se and Bi–Se Alloys
(72.15; 61.25; 75.20; S4)
(b) Vol. **140**, No. 1, p. 233—242 (1987).

W. ZACHARKO
Experimental Investigations on the Anisotropy of the Upper Critical Field in Hexagonal Intrinsic Type II Superconductor $InSn_4$
(74.70; S4)
(b) Vol. **140**, No. 1, p. K33—K36 (1987).

D. A. AKSENOV, B. M. BOLOTOV, A. I. POLYAKOV, and H. V. CHASNIKOVA
On the Change of Dipolar Shape of the NMR Line by Quadrupolar Interaction
(76.60; S4)
(b) Vol. **140**, No. 1, p. K51—K54 (1987).

A. KUMAR, G. CHANDRA, and O. P. KATYAL
The Effect of Grain Boundary Scattering on the Electrical Resistivity of Polycrystalline Lead, Tin, and Tin–Lead Alloy Films
(72.15 and 73.60; 78.65; S4)
(a) Vol. **100**, No. 1, p. 169—175 (1987).

TRAN THOAI DUY BAO
Image Force and Stopping Power of a Fast Moving Ion near a Metal Surface
(61.80 and 73.20; 71.45; S4)
(b) Vol. **140**, No. 2, p. 449—457 (1987).

L. K. MISRA and D. N. TRIPATHY
Estimation of Plasmon Dispersion-Coefficient and Evaluation of the Effective Mass of Electrons in Metals
(71.45; 71.25; S2; S4)
(b) Vol. **140**, No. 2, p. 477—490 (1987).

Xian-Wu Zou, Zhun-Zhi Jin, and Sheng-Qun Tang
Temperature Dependence of the Static Structure Factor for Liquid Bi, Ga, Sb, and Sn
(61.25; S4)
(b) Vol. **142**, No. 1, p. 9—14 (1987).

V. Ramamurthy
On the Inconsistencies of Axially Symmetric Force Models for Noncubic Solids
(63.20; 62.20; S4)
(b) Vol. **142**, No. 1, p. 67—76 (1987).

H.-E. Schaefer
Investigation of Thermal Equilibrium Vacancies in Metals by Positron Annihilation
(61.70; 71.60; 78.70; S1; S4)
(a) Vol. **102**, No. 1, p. 47—65 (1987).

A. Belaidi and R. N. West
The Effects of Lattice Anisotropy on Positron Parameters in In–Pb Alloys
(61.55 and 71.60; 78.70; S4)
(a) Vol. **102**, No. 1, p. 107—112 (1987).

A. Bharathi, C. S. Sundar, D. Vasumathi, and K. P. Gopinathan
A Positron Annihilation Study of Clustering of Ag Atoms in a Quenched Al–1 at% Ag Alloy
(64.75 and 78.70; 61.70; S4)
(a) Vol. **102**, No. 1, p. 139—143 (1987).

M. Eldrup and K. O. Jensen
Positron Trapping Rates into Cavities in Al: Temperature and Size Effects
(61.70 and 78.70; S4)
(a) Vol. **102**, No. 1, p. 145—152 (1987).

D. O. Northwood and I. O. Smith
The Creep Behaviour of a Neutron Irradiated Al–0.3 wt% Li Alloy at 300 °C
(61.70 and 62.20; S4)
(a) Vol. **102**, No. 1, p. 251—263 (1987).

T. H. Youssef, F. A. Saadalah, and I. K. Bishay
Plastic Behaviour of Polycrystalline Al–0.72 wt% Si
(62.20; S4)
(a) Vol. **102**, No. 1, p. 273—278 (1987).

E. S. M. Higgy, S. M. Youssef, S. A. El-Sahhar, and H. H. Talaat
Electrical Properties of Thin Polycrystalline Lead Films
(73.60; S4)
(a) Vol. **102**, No. 1, p. 301—306 (1987).

N. B. Brandt, M. Yu. Lavrenyuk, N. Ya. Minina, A. M. Savin, W. Kraak, and R. Herrmann
Thermal E.M.F. Anomalies Due to Axial Compression and the Band Structure of $Bi_{1-x}Sb_x$ ($x = 0.27$) Alloys
(71.25 and 72.15; S4)
(b) Vol. **143**, No. 2, p. 601—609 (1987).

Cs. Szeles, K. Süvegh, Z. Homonnay, and A. Vértes
Positron Lifetime and Mössbauer Spectroscopy Study of Vacancy–Tin Interaction in Dilute Al–Sn Alloys
(61.70; 76.80; 78.70; S4)
(a) Vol. **103**, No. 2, p. 397—401 (1987).

I. Kanazawa, H. Murakami, and M. Doyama
The Migration Stage of Divacancies in Quenched Pure Aluminium, Dilute Aluminium Alloys, and Pure Copper Studied by Positron Annihilation
(61.70; 78.70; S1.2; S4)
(a) Vol. **103**, No. 2, p. 403—411 (1987).

T. S. Kê, Q. Tan, and Q. F. Fang
Further Experiments on the Anomalously Amplitude-Dependent Internal Friction Peaks in Polycrystalline and Single-Crystal Al–Mg
(62.40; S4)
(a) Vol. **103**, No. 2, p. 421—429 (1987).

S. Ohtaki
The Nature of Stage II_A Recovery in Cold-Worked Aluminium
(62.20; 61.70; 72.15; S4)
(a) Vol. **103**, No. 2, p. 449—457 (1987).

J. Sosnowski, M. Ciszek, A. J. Zaleski, and E. Trojnar
Analysis of Losses Measurements on Cylindrical Pb–In Samples through Flux Cutting Effect
(74.70; S4)
(a) Vol. **103**, No. 2, p. 527—530 (1987).

M. Ohta, T. Kanadani, A. Sakakibara, and M. Yamada
On the Two Populations of GP Zones in the Al–Zn Alloy Aged after Two-Step Quenching
(62.20; S4)
(a) Vol. **103**, No. 2, p. K85—K86 (1987).

H. Bross and R. Eder
Self-Consistent MAPW Calculation with a Warped Muffin-Tin Potential. I. The Electronic Structure of Al and Its Pressure Dependence
(71.20 and 71.25; 64.30; S4)
(b) Vol. **144**, No. 1, p. 175—193 (1987).

J. N. Mundy
Diffusion Mechanism in F.C.C. Metals
(66.30; S4)
(b) Vol. **144**, No. 1, p. 233—241 (1987).

G. J. Vázquez and L. F. Magaña
Ab-initio Calculation of the Volume Dependence of the Resistivity of Aluminium
(72.15; 63.20; S4)
(b) Vol. **144**, No. 2, p. K119—K123 (1987).

T. Tsuchida and R. Kaneko
Temperature Dependence of Activation Energy in Stage II Recovery in Deformed Aluminium
(61.70; 72.15; S4)
(a) Vol. **104**, No. 2, p. 627—634 (1987).

E. Borchi, S. De Gennaro, L. Lombardini, and M. Zoli
Elastic Constants and Young's Modulus for Texturized Materials with Cubic and Hexagonal Symmetries
(62.20; S1.2; S4)
(a) Vol. **104**, No. 2, p. 667—678 (1987).

S. C. Yan and T. S. Kê
Internal Friction Peaks Associated with the Polygonization Boundaries in Aluminium and Dilute Aluminium–Copper Alloys
(62.40; S4)
(a) Vol. **104**, No. 2, p. 715—721 (1987).

Q. Tan and T. S. Kê
Double Amplitude Internal Friction Peaks in Al–0.02 wt% Mg Single Crystals
(62.40; S4)
(a) Vol. **104**, No. 2, p. 723—729 (1987).

X. H. Wang, Y. B. Xu, Z. G. Wang, and K. F. Ha
The Observation of the Dislocation-Free Zone ahead of the Crack-Tip in Bulk Aluminium Single Crystal
(61.70 and 62.20; S4)
(a) Vol. **104**, No. 2, p. K79—K82 (1987).

S5. Nonmetallic elements

O. A. Golikova
Semiconductors with Complex Lattice and the Amorphization Problem **(Review Article)**
(61.40 and 72.20; S5)
(a) Vol. **101**, No. 2, p. 277—314 (1987).

V. K. Bashenov, D. L. Kardashev, and D. I. Marvakov
Orbital Removal Method for the Neutral Vacancy in Semiconductors
(71.55; S5)
(b) Vol. **139**, No. 1, p. K31—K35 (1987).

M. Moussavi-Madani and P. C. Rice-Evans
Search for Positron Trapping in Trigonal Selenium
(61.70 and 78.70; S5)
(b) Vol. **139**, No. 2, p. K99—K102 (1987).

A. Zaiour, M. Hage-Ali, J. M. Koebel, A. Bentz, and P. Siefert
Determination of Segregation Coefficients of a Series of Impurities in Tellurium
(61.70 and 64.75; S5)
(a) Vol. **100**, No. 2, p. K139—K143 (1987).

S. Orzeszko, K. Fabisiak, and F. Rozpłoch
Optical and Mobility (Electrical) Gaps of Arc Evaporated Hydrogenated Amorphous Carbon Films
(73.60 and 78.65; S5)
(a) Vol. **101**, No. 1, p. 159—163 (1987).

S. Shimamura, K. Shindo, and A. Morita
Structural Stability and Phonon Dispersion of Arsenic
(63.20; S5)
(b) Vol. **142**, No. 1, p. 89—96 (1987).

Deyan He, Fangqing Zhang, Jingzhong Cui, and Guanghua Chen
Interface Defects in Sputtered a-Si:H/a-C:H Multilayers
(73.40; 76.30; 79.20; S5; S5.11)
(b) Vol. **142**, No. 1, p. K35—K37 (1987).

A. Seeger and F. Banhart
On the Systematics of Positron Lifetimes in Metals
(71.60; 61.70; S2; S5)
(a) Vol. **102**, No. 1, p. 171—179 (1987).

L. Cser, M. Kocsis, L. Köszegi, A. Oles, A. Topolnicka, and J. Wolny
Neutron Scattering Study of Carbon Fibres
(61.55; 61.12; S5)
(a) Vol. **103**, No. 1, p. 39—44 (1987).

A. K. Abass
Optical Properties of α-Sulfur Thin Films
(78.65; S5)
(a) Vol. **103**, No. 1, p. 281—284 (1987).

C. O. Rodríguez, R. A. Casali, E. L. Peltzer y Blancá, and O. M. Cappannini
First Principles Prediction of Structural Properties and Pressure Dependence of the Charge Density and Energy Gaps in Diamond
(61.50 and 71.45; 71.25; S5)
(b) Vol. **143**, No. 2, p. 539—548 (1987).

E. G. Vyatkin and S. A. Vorobiev
Molecular-Type Motion of Relativistic Electrons at Planar Channeling in Crystals
(61.80; S5)
(b) Vol. **144**, No. 2, p. 487—499 (1987).

S5.11. Silicon

A. A. Karpushin, A. N. Sorokin, I. Yu. Semenova, and M. Tomášek
A Contribution to the Theoretical Interpretation of the UPS Spectra of Monoatomic Adsorption of Oxygen on the Si (111) Surface
(68.20 and 73.20; 79.60; S5.11)
(b) Vol. **139**, No. 1, p. 125—130 (1987).

V. I. Sokolov and N. A. Fedorovich
Mechanical Stresses on the Si–SiO_2 Interface
(68.25; S5.11; S10.1)
(a) Vol. **99**, No. 1, p. 151—158 (1987).

Zhang Fangqing, Xu Xixiang, and Chen Guanghua
Persistent Photoconductivity in a-Si:H/a-$Si_{1-x}C_x$:H Multilayers
(73.60; 72.40; S5.11; S6)
(a) Vol. **99**, No. 1, p. 165—170 (1987).

J.-P. Zöllner, G. Paasch, and H. Übensee
The Influence of Quantum Effects on the Semiconductor lf- and hf-Capacitance
(73.40; S5.11)
(a) Vol. **99**, No. 1, p. 193—204 (1987).

H. Lemke
Doping Properties of Nickel in Silicon
(71.55; S5.11)
(a) Vol. **99**, No. 1, p. 205—213 (1987).

H. K. Phan, L. H. Phu, and N. T. Nguyet Anh
Switching Characteristics of MISS Devices Containing Electron Surface Traps
(73.40 and 85; S5.11)
(a) Vol. **99**, No. 1, p. 309—314 (1987).

K. SCHMALZ, K. TITTELBACH-HELMRICH, and H. RICHTER
On DLTS-Measurements of the Barrier-Limited Capture Process
(71.55 and 73.30; S5.11)
(a) Vol. **99**, No. 2, p. K97—K102 (1987).

N. T. TRAN and F. R. JEFFREY
Effects of Carbon Profiles at the p/i Interface in Amorphous Silicon Solar Cells
(73.40 and 85; S5.11; S6)
(a) Vol. **99**, No. 2, p. K143—K148 (1987).

V. I. GRIDNEV, V. G. KHLABUTIN, E. I. ROZUM, and S. A. VOROBIEV
Observation of Similarity of Angular Distributions of Electrons at Planar Channeling in Silicon Crystal
(61.80; S5.11)
(b) Vol. **140**, No. 1, p. K13—K17 (1987).

T. DRÜSEDAU, D. WEGENER, and R. BINDEMANN
The Ordinary and Generalized Meyer-Neldel Rule in a-Si:H Calculated by the Statistical Fermi Level Shift
(71.20 and 72.20; S5.11)
(b) Vol. **140**, No. 1, p. K27—K31 (1987).

R. GERMER
X-Ray Topography with a Semiconductor TV System
(61.10; S5.11)
(a) Vol. **100**, No. 1, p. 47—51 (1987).

K. SCHMALZ, F.-G. KIRSCHT, S. NIESE, H. RICHTER, M. KITTLER, W. SEIFERT, I. BABANSKAYA, H. KLOSE, K. TITTELBACH-HELMRICH, and J. SCHÖNEICH
On the Intrinsic Gettering in Fe-Contaminated Cz-Si
(61.70; 71.55; S5.11)
(a) Vol. **100**, No. 1, p. 69—85 (1987).

P. ZAUMSEIL, U. WINTER, F. CEMBALI, M. SERVIDORI, and Z. SOUREK
Determination of Dislocation Loop Size and Density in Ion Implanted and Annealed Silicon by Simulation of Triple Crystal X-Ray Rocking Curves
(61.70; 61.10; S5.11)
(a) Vol. **100**, No. 1, p. 95—104 (1987).

L. ŠAMAJ
Recombination Processes at Grain Boundaries in Polycrystalline Semiconductors
(73.40; S5.11)
(a) Vol. **100**, No. 1, p. 157—167 (1987).

G. F. CEROFOLINI and M. L. POLIGNANO
Self-Interstitials and Generation Lifetime in Silicon p–n Junctions
(73.40; 61.70; S5.11)
(a) Vol. **100**, No. 1, p. 177—186 (1987).

B. BALLAND, A. GLACHANT, A. RONDA, and J. C. DUPUY
High-Temperature Thermal Nitridation and Low-Temperature Electron-Beam-Enhanced Nitridation of SiO_2 Thin Films
(68.48 and 73.40; S5.11; S10.1)
(a) Vol. **100**, No. 1, p. 187—198 (1987).

M. Kohyama
Structures and Energies of Symmetrical ⟨001⟩ Tilt Grain Boundaries in Silicon
(61.70; S5.11)
(b) Vol. **141**, No. 1, p. 71—83 (1987).

A. Latgé, S. S. Makler, and E. V. Anda
Electronic Properties of Amorphous Semiconductors. Lattice Topology Effects
(61.40 and 71.20; S5.11)
(b) Vol. **141**, No. 1, p. 163—171 (1987).

S. K. Savvinykh
On the Theory of "Deep" Electronic Impurity States in Many-Valley Semiconductors
(71.55; S5.11; S5.12)
(b) Vol. **141**, No. 1, p. 185—190 (1987).

W. Cordts, H. Wiesner, and J. Monecke
Pressure Dependence of the Electron Density in Silicon
(71.20; S5.11)
(b) Vol. **141**, No. 1, p. K53—K56 (1987).

M. Takeda
A Simplified Method for Electron Dynamical Diffraction with Crystal Symmetry Considerations
(61.14; S4; S5.11)
(a) Vol. **101**, No. 1, p. 25—36 (1987).

V. E. Borisenko and S. G. Yudin
Steady-State Solubility of Substitutional Impurities in Silicon
(64.75; S5.11)
(a) Vol. **101**, No. 1, p. 123—127 (1987).

L. Šamaj
The Lifetime of Minority Carriers in Polycrystalline Semiconductors
(72.20 and 73.40; S5.11)
(a) Vol. **101**, No. 1, p. 137—141 (1987).

H. Lemke
Impurity Reactions in the Surface Regions of Zn and Pt Doped Si Crystals
(61.70 and 71.55; S5.11)
(a) Vol. **101**, No. 1, p. 193—203 (1987).

I. A. Abroyan, M. V. Kruglov, and I. K. Solomin
Surface-Layer Swelling on Implantation of Hydrogen Ions into Silicon at Different Doping Concentrations
(61.80; 68.20; S5.11)
(a) Vol. **101**, No. 1, p. K5—K8 (1987).

N. Balasubramanyam and Vikram Kumar
Pressure Dependence of Barrier Heights of Schottky Contacts on Silicon
(73.30; S5.11)
(a) Vol. **101**, No. 1, p. K29—K32 (1987).

H.-Matsuo Kagaya, N. Shoji, and T. Soma
Specific Heat and Thermal Expansion at High Temperatures of Si and Ge
(65.40 and 65.70; S5.11; S5.12)
(b) Vol. **142**, No. 1, p. K13—K17 (1987).

Deyan He, Fangqing Zhang, Jingzhong Cui, and Guanghua Chen
Interface Defects in Sputtered a-Si:H/a-C:H Multilayers
(73.40; 76.30; 79.20; S5; S5.11)
(b) Vol. **142**, No. 1, p. K35—K37 (1987).

R. Würschum, M. Scheytt, and H.-E. Schaefer
Nanocrystalline Metals and Semiconductors Studied by Positron Lifetime Spectroscopy
(61.55 and 78.70; 61.70; S1.1; S1.2; S1.3; S5.11)
(a) Vol. **102**, No. 1, p. 119—126 (1987).

V. A. Labunov, V. P. Bondarenko, V. E. Borisenko, and A. M. Dorofeev
High-Temperature Treatment of Porous Silicon
(61.55; S5.11)
(a) Vol. **102**, No. 1, p. 193—198 (1987).

G. A. Kachurin, I. E. Tyshchenko, L. I. Fedina, E. Wieser, and Ch. Weise
High Temperature Implantation of Silicon by B^+ and BF_2^+ Ions
(61.80; 66.30; S5.11)
(a) Vol. **102**, No. 1, p. 265—272 (1987).

G. F. Cerofolini
A Model for the Non-Equilibrium, Pure-Generation Centre in Silicon
(71.55; 73.40; S5.11)
(a) Vol. **102**, No. 1, p. 345—348 (1987).

C. Donolato and M. Bianconi
Use of the EBIC Contrast Profile Area for Evaluating the Recombination Strength of Dislocations
(61.70; 72.20; S5.11)
(a) Vol. **102**, No. 1, p. K7—K11 (1987).

A. M. Mazzone
Monte Carlo Simulation of Channeled and Random Profiles of Heavy Ions Implanted in Silicon at High Energy (1.2 MeV)
(61.70; 61.80; S5.11)
(a) Vol. **102**, No. 1, p. K13—K18 (1987).

H. Kitagawa and H. Nakashima
Nickel-Related Donor Level in Silicon
(71.55; 72.80; S5.11)
(a) Vol. **102**, No. 1, p. K23—K27 (1987).

V. I. Gatalskaya, E. M. Gololobov, and L. A. Kurochkin
Annealing Effect on Charge Carrier Scattering in Irradiated Silicon from Cyclotron Resonance Data
(61.80 and 76.40; S5.11)
(b) Vol. **142**, No. 2, p. K105—K108 (1987).

A. F. Khokhlov, A. I. Mashin, A. V. Ershov, Yu. A. Mordvinova, and D. A. Pavlov
The Properties of Amorphous Silicon Doped with Isovalent Impurities
(73.60 and 78.65; 61.70; S5.11)
(b) Vol. **142**, No. 2, p. K125—K129 (1987).

F. LUKEŠ
Temperature Dependence of Ellipsometric Angles of Silicon
(78.20; 68.45; S5.11)
(a) Vol. **102**, No. 2, p. 803—814 (1987).

D. I. TSIULYANU, E. P. KOLOMEYKO, and V. N. STAMOV
Photoelectric Phenomena in Isotope Vitreous As_2S_3–Silicon Single-Crystal Heterojunctions
(72.40 and 73.40; S5.11; S8)
(a) Vol. **102**, No. 2, p. K103—K107 (1987).

CH. SCHNITTLER and S. SCHERF
Alternative Quasi-Classical Transport Equations. II. Boltzmann's Equation and the Electron Distribution Function for MIS Systems
(73.40; 73.25; S5.11)
(b) Vol. **143**, No. 1, p. 325—333 (1987).

A. I. KASIYAN, A. G. CHEBAN, M. A. CHEBAN, I. V. SUR, and O. V. ZAKHARCHUK
Effect of Traps on the Shape of the Output Signal of Ionizing-Radiation Semiconductor Defectors
(72.20 and 85; S5.11)
(a) Vol. **103**, No. 1, p. 185—191 (1987).

H. BERGNER, V. BRÜCKNER, F. KERSTAN, M. LENZNER, and W. NOWICK
An Optoelectronic Switch with Increased Sensitivity Using a Finger Structure Electrode
(72.40; S5.11)
(a) Vol. **103**, No. 1, p. 193—197 (1987).

D. P. JOSHI
Grain Boundary Recombination in Polycrystalline Silicon under Optical Illumination
(72.40; 73.40; S5.11)
(a) Vol. **103**, No. 1, p. 213—218 (1987).

M. L. POLIGNANO, G. F. CEROFOLINI, H. BENDER, C. CLAEYS, and J. REFFLE
The Role of Oxygen in Silicon p–n Junction Gettering
(61.70 and 73.40; S5.11)
(a) Vol. **103**, No. 1, p. 307—316 (1987).

A. STESMANS
Interaction between Localized and Delocalized States Indicated from Temperature-Dependent ESR Study of Submetallic P-Doped Si
(76.30; 71.25; S5.11)
(b) Vol. **143**, No. 2, p. 733—740 (1987).

S. SCHERF, CH. SCHNITTLER, and G. PAASCH
Alternative Quasi-Classical Transport Equations. III. Charge Transport in MIS Systems with Inclusion of Diffuse Surface Scattering and Quantum Size Effects
(73.40; 73.25; S5.11)
(b) Vol. **143**, No. 2, p. 795—805 (1987).

U. PIETSCH and K. UNGER
An Experimental Proof of the Valence Electron Density Variation in Silicon under High Electric Field
(61.10 and 61.50; S5.11)
(b) Vol. **143**, No. 2, p. K95—K97 (1987).

H. Mehrer
Implantation of Radioisotopes at ISOLDE. A Novel Source Production Technique for Diffusion Studies in Solids
(66.30; 61.70; S1.2; S1.3; S5.11)
(a) Vol. **104**, No. 1, p. 247—261 (1987).

T. I. Galkina, A. Yu. Blinov, M. M. Bonch-Osmolovskii, P. N. Lebedev, O. Koblinger, K. Lassmann, and W. Eisenmenger
Down-Conversion of High-Frequency Acoustic Phonons in a-Si:H
(63.20; S5.11)
(b) Vol. **144**, No. 2, p. K87—K91 (1987).

A. B. Gerasimov, Z. G. Gogua, D. A. Kimeridze, and V. N. Svanidze
The Role of the Central Cell in the Problem of Carrier Capture by Attractive Centres in Ge and Si
(71.55; S5.11; S5.12)
(b) Vol. **144**, No. 2, p. K99—K104 (1987).

A. Jakubowicz, H.-U. Habermeier, A. Eisenbeiss, and D. Käss
Electron Beam Induced Current versus Temperature Investigations of Localized Dislocations in Heat-Treated Czochralski Silicon
(61.70; S5.11)
(a) Vol. **104**, No. 2, p. 635—641 (1987).

R. C. Martin and N. M. Ghoniem
Monte Carlo Simulation of Coupled Ion–Electron Transport in Semiconductors
(61.80 and 73.40; 79.20; S5.11)
(a) Vol. **104**, No. 2, p. 743—754 (1987).

A. N. Isotov and E. A. Steinman
Reconstruction of Optical Dislocation Centres under the Action of Shear Stresses
(61.70 and 71.55; 78.55; S5.11)
(a) Vol. **104**, No. 2, p. 777—784 (1987).

G. Ghibaudo
Calculation of Surface Charge Noise at the Si–SiO_2 Interface
(73.20 and 73.40; S5.11)
(a) Vol. **104**, No. 2, p. 917—930 (1987).

A. M. Afanasev, R. M. Imamov, E. Kh. Mukhamedzhanov, and A. A. Nazlukhanyan
Coherent Compton Effect in the Grazing Bragg-Laue Geometry
(61.10; S5.11)
(a) Vol. **104**, No. 2, p. K73—K77 (1987).

H. Johansen and B. Lämmel
Optical Feedback during CW Laser Annealing of Silicon Detected by SEM-EBIC
(61.80; 79.20; S5.11)
(a) Vol. **104**, No. 2, p. K83—K84 (1987).

S5.12. Germanium

H.-Matsuo Kagaya, N. Shoji, and T. Soma
Specific Heat and Thermal Expansion at High Temperatures of Si and Ge
(65.40 and 65.70; S5.11; S5.12)
(b) Vol. **142**, No. 1, p. K13—K17 (1987).

I. G. Kirnas, P. G. Litovchenko, E. E. Petrosyan, and N. N. Pashchuk
Calculation of Bulk Charge and Electric Field Profiles in One-Open-Face Coaxial γ-Detectors Using Experimental *C–U* Characteristics
(73.40; S5.12)
(a) Vol. **102**, No. 1, p. 315—319 (1987).

R. A. Evarestov and V. P. Smirnov
Symmetry of Localized Crystalline Orbitals and Double-Valued Band Representation
(71.50; S5.12)
(b) Vol. **142**, No. 2, p. 493—499 (1987).

I. Dézsi, H. Engelmann, U. Gonser, and G. Langouche
Cobalt–Germanium Structures Investigated by Mössbauer Spectroscopy
(61.55; 61.80; 76.80; S1.61; S5.12)
(a) Vol. **104**, No. 1, p. 417—426 (1987).

Yi-ben Xia, E. Bangert, and G. Landwehr
Low-Temperature Magnetoresistance in Medium-Angle Grain Boundaries in Germanium
(73.40; S5.12)
(b) Vol. **144**, No. 2, p. 601—608 (1987).

A. B. Gerasimov, Z. G. Gogua, D. A. Kimeridze, and V. N. Svanidze
The Role of the Central Cell in the Problem of Carrier Capture by Attractive Centres in Ge and Si
(71.55; S5.11; S5.12)
(b) Vol. **144**, No. 2, p. K99—K104 (1987).

S5.2. Solidified Gases

S6. Group IV compounds
(for transition metal compounds, see S1.61)

J. Bullot and M. P. Schmidt
Physics of Amorphous Silicon–Carbon Alloys **(Review Article)**
(61.40; 71.20; 71.55; 72.40; 73.60; 78.20; 78.55; 78.60; S6)
(b) Vol. **143**, No. 2, p. 345—418 (1987).

H. Rübel, B. Schröder, W. Fuhs, J. Krauskopf, T. Rupp, and K. Bethge
IR Spectroscopy and Structure of RF Magnetron Sputtered a-SiC:H Films
(68.55 and 73.60; 78.30; S6)
(b) Vol. **139**, No. 1, p. 131—143 (1987).

Zhang Fangqing, Xu Xixiang, and Chen Guanghua
Persistent Photoconductivity in a-Si:H/a-$Si_{1-x}C_x$:H Multilayers
(73.60; 72.40; S5.11; S6)
(a) Vol. **99**, No. 1, p. 165—170 (1987).

N. T. Tran and F. R. Jeffrey
Effects of Carbon Profiles at the p/i Interface in Amorphous Silicon Solar Cells
(73.40 and 85; S5.11; S6)
(a) Vol. **99**, No. 2, p. K143—K148 (1987).

W. Siebert, R. Carius, W. Fuhs, and K. Jahn
Photoluminescence in a-$Si_{1-x}C_x$:H Films
(78.55; 78.65; S6)
(b) Vol. **140**, No. 1, p. 311—321 (1987).

H. Yamamoto
Resonant Tunneling through ABCBA-Type Double-Barrier Structures
(73.40; S5.11; S6; S7.12)
(b) Vol. **141**, No. 2, p. K179—K183 (1987).

M. T. Sebastian, K. Narayanan, and P. Krishna
X-Ray Diffraction Effects from 2H Crystals Undergoing Transformation to the 3C Structure by the Layer Displacement Mechanism
(61.10; 61.70; 64.70; S6)
(a) Vol. **102**, No. 1, p. 241—249 (1987).

H.-Matsuo Kagaya and T. Soma
The Influence of the Compression Effect on the Phase Diagrams of Si–Ge System
(62.50; 64.75; 65.50; S6)
(b) Vol. **144**, No. 2, p. 549—556 (1987).

S7. Group V compounds
(for transition metal compounds, see S1.62)

S7.1. III–V compounds

A. P. Dmitriev, M. P. Mikhailova, and I. N. Yassievich
Impact Ionization in $A^{III}B^{V}$ Semiconductors in High Electric Fields **(Review Article)**
(79.20; S7.1)
(b) Vol. **140**, No. 1, p. 9—37 (1987).

G. Dlubek and R. Krause
Positron Studies of Defects in III–V Semiconductor Compounds **(Review Article)**
(61.70 and 78.70; S7.1)
(a) Vol. **102**, No. 2, p. 443—479 (1987).

T. Jung and M. Schmidt
Plasma-Assisted Reactive Evaporation of Aluminium Nitride Films
(68.55 and 73.60; 79.20; S7.1)
(a) Vol. **100**, No. 1, p. 207—211 (1987).

L. Wendler and R. Haupt
Electron–Phonon Interaction in Semi-Infinite Semiconductor Superlattices
(71.38 and 71.45; 63.20; 68.30; 73.40; S7.1; S7.15)
(b) Vol. **141**, No. 2, p. 493—505 (1987).

W. Kühn, R. Strehlow, and M. Hanke
Self-Consistent Tight-Binding Investigation of Chemical Trends for Native Defects in III–V Semiconductors
(71.55; S7.1)
(b) Vol. **141**, No. 2, p. 541—557 (1987).

H.-Matsuo Kagaya and T. Soma
Mode Grüneisen Parameters and Pressure-Induced Phase Transition of Tetrahedral Compounds
(63.20; 62.20; 64.70; S7.1)
(b) Vol. **142**, No. 1, p. 97—103 (1987).

H.-Matsuo Kagaya and T. Soma
Thermal Properties of Tetrahedral Compounds
(65.40 and 65.70; S7.1; S8.1)
(b) Vol. **142**, No. 2, p. 411—416 (1987).

J. C. Bourgoin, H. J. von Bardleben, and D. Stievenard
Irradiation Induced Defects in III–V Semiconductor Compounds
(61.80; 61.70; 78.70; S7.1)
(a) Vol. **102**, No. 2, p. 499—510 (1987).

A. I. Lukomskii, V. B. Shipilo, E. M. Shishonok, and N. G. Anichenko
Raman Scattering of Cubic Boron Nitride
(78.30; S7.1)
(a) Vol. **102**, No. 2, p. K137—K139 (1987).

L. Wendler and R. Haupt
Electron–Phonon Interaction in Semiconductor Superlattices
(71.38; 63.20; 68.30; 73.40; S7.1; S7.15)
(b) Vol. **143**, No. 2, p. 487—510 (1987).

A. V. Dobrynin, G. A. Naida, and V. A. Novoselov
Acoustoelectric Parameters of Aluminum Nitride Epitaxial Layers
(72.50; S7.1)
(a) Vol. **104**, No. 1, p. K47—K51 (1987).

Ch. Khamsasy, Nguyen Van Hieu, and Nguyen Ai Viet
Three-Branch Excitonic Polaritons in Direct Band Gap Cubic Semiconductors
(71.35 and 71.36; S7.1)
(b) Vol. **144**, No. 2, p. 645—652 (1987).

S7.11. Phosphides

J. S. Nkoma
Surface Phonon Polaritons in Semi-Infinite Semiconductor Superlattices
(68.30 and 71.36; S7.11; S7.12)
(b) Vol. **139**, No. 1, p. 117—123 (1987).

T. Soma and H.-Matsuo Kagaya
Compression Effect on Elastic Coefficients of AlP, AlAs, and AlSb Compounds
(62.20; S7.11; S7.12; S7.13)
(b) Vol. **139**, No. 1, p. K1—K4 (1987).

M. S. El-Dessouki, V. A. Attia, M. M. Saad El-Deen, and F. Z. Gobrial
Surface Photovoltage Spectroscopy of Real ⟨111⟩ GaP Surfaces
(73.20 and 73.25; S7.11)
(a) Vol. **99**, No. 1, p. 171—179 (1987).

M. Zafar Iqbal, A. Jabbar, N. Baber, and N. Zafar
An Inadvertent Mid-Gap Electron Level in Liquid-Phase-Epitaxial Gap
(71.55 and 85; S7.11)
(a) Vol. **99**, No. 1, p. K65—K68 (1987).

G. Irmer, J. Monecke, B. H. Bairamov, and V. V. Toporov
On the Electronic Raman Scattering at Minority Impurities in GaP
(71.55 and 78.30; S7.11)
(b) Vol. **139**, No. 2, p. K157—K160 (1987).

O. GAZAKOV, A. KH. ORAZBERDYEV, and YA. CHARYEV
Photoelectric Properties of the M–n-In_2O_3–(n, p)-InP Surface Barrier Structures
(72.40 and 73.40; S7.11)
(a) Vol. **100**, No. 1, p. 139—142 (1987).

W. ULRICI, L. EAVES, K. FRIEDLAND, and D. P. HALLIDAY
Optical Studies of Vanadium in Gallium Phosphide
(71.55; 72.80; 78.50; 78.55; S7.11)
(b) Vol. **141**, No. 1, p. 191—202 (1987).

B. BOUCHIKHI, C. MICHEL, S. RAVELET, and B. LEPLEY
Preparation and Electrical Properties of Thin Native Oxide Double-Layer Insulator Films on n-Type InP
(68.55 and 73.40; S7.11)
(a) Vol. **101**, No. 1, p. 173—184 (1987).

H. SIETHOFF, J. VÖLKL, D. GERTHSEN, and H. G. BRION
The Lower Yield Point of InP and GaAs
(62.20; S7.11; S7.12)
(a) Vol. **101**, No. 1, p. K13—K18 (1987).

M. GARRIGUES and S. K. KRAWCZYK
Erratum to: Characterization of the InP Surface by Photoluminescence Imaging
(78.55; S7.11)
(a) Vol. **101**, No. 2, p. K181 (1987).

J. MONECKE, W. CORDTS, G. IRMER, B. H. BAIRAMOV, and V. V. TOPOROV
LO-Phonon Modes Bound to Neutral Impurities in Planar Semiconductors. II. Special Cases and Experimental Results
(63.20 and 71.55; 71.38; S7.11)
(b) Vol. **142**, No. 1, p. 237—246 (1987).

BIZHONG ZHOU, JIANGLING FONG, and JINGZHAO HUANG
Investigation of Defects in GaP Using Positron Annihilation
(61.70 and 78.70; S7.11)
(a) Vol. **102**, No. 2, p. 533—536 (1987).

C. D. BELING, R. I. SIMPSON, M. G. STEWART, Y. Y. WANG, S. FUNG, J. C. H. WAI, and T. N. SUN
Positron Mobility Measurements and Their Relevance to Defect and Impurity Studies in Semiconductors and Insulators
(61.70 and 78.70; S5.11; S7.11)
(a) Vol. **102**, No. 2, p. 537—547 (1987).

J. KREISSL, W. GEHLHOFF, and W. ULRICI
Electron Paramagnetic Resonance of Titanium in GaP
(71.55 and 76.30; S7.11)
(b) Vol. **143**, No. 1, p. 207—215 (1987).

U. Woggon, F. Henneberger, and H. Weinert
On the Bleaching of the A-Bound Exciton Absorption in GaP: N. An Example for an Electron–Hole Liquid Induced Optical Nonlinearity
(71.35 and 78.55; S7.11)
(b) Vol. **143**, No. 2, p. 611—623 (1987).

B. Ulrici, R. Stedman, and W. Ulrici
Local Vibrational Mode Absorption of Hydrogen and Oxygen Centres in LEC-Grown GaP and GaAs
(63.20 and 78.30; 61.70; S7.11; S7.12)
(b) Vol. **143**, No. 2, p. K135—K139 (1987).

P. Deus, H. A. Schneider, U. Voland, and K. Stiehler
Low Temperature Thermal Expansion of InP
(65.70; S7.11)
(a) Vol. **103**, No. 2, p. 443—447 (1987).

C. Ascheron, H. Neumann, V. N. Bugrov, and S. A. Karamyan
Microhardness of Ion Implanted GaP
(61.70 and 62.20; S7.11)
(a) Vol. **103**, No. 2, p. K81—K84 (1987).

A. Haufe
Nitrogen Bound Electrons, Excitons, and Multiexcitons in GaP
(71.35 and 71.55; S7.11)
(b) Vol. **144**, No. 2, p. 733—738 (1987).

S7.12. Arsenides

J. S. Nkoma
Surface Phonon Polaritons in Semi-Infinite Semiconductor Superlattices
(68.30 and 71.36; S7.11; S7.12)
(b) Vol. **139**, No. 1, p. 117—123 (1987).

G. Manzke, K. Henneberger, and V. May
Many-Exciton Theory for Multiple Quantum-Well Structures
(71.35; 78.20; S7.12; S7.15)
(b) Vol. **139**, No. 1, p. 233—239 (1987).

T. Soma and H.-Matsuo Kagaya
Compression Effect on Elastic Coefficients of AlP, AlAs, and AlSb Compounds
(62.20; S7.11; S7.12; S7.13)
(b) Vol. **139**, No. 1, p. K1—K4 (1987).

YU. B. BOLKHOVITYANOV and R. I. BOLKHOVITYANOVA
A GaAs Surface after Its Contact with an Undersaturated Al–Ga–As Liquid. Observation after Anodic Oxidation
(68.45; S7.12)
(a) Vol. **99**, No. 1, p. 159—164 (1987).

P. ENDER
Minority-Carrier Leakage over Step p–n Heterojunctions
(73.40; S7.12)
(a) Vol. **99**, No. 1, p. 181—192 (1987).

O. BREITENSTEIN and L. J. GILING
Scanning-DLTS Investigations on Semi-Insulating GaAs:Cr, In Containing "Streamers"
(71.55; S7.12)
(a) Vol. **99**, No. 1, p. 215—223 (1987).

J. PASTRŇÁK, F. KAREL, J. OSWALD, W. ULRICI, and N. M. KOLCHANOVA
The Study of Impurity Levels in LEC Semi-Insulating GaAs:Cr Samples by Two Source Photoconductivity Measurements
(71.55; 72.40; 78.55; S7.12)
(a) Vol. **99**, No. 1, p. 225—235 (1987).

F. COMAS, C. TRALLERO GINER, and J. TUTOR
Ohmic Mobility of Quasi-One-Dimensional Semiconducting Structures. Acoustic and Piezoelectric Phonon Scattering
(72.20 and 73.60; S7.12; S8.11)
(b) Vol. **139**, No. 2, p. 433—440 (1987).

A. LEFEBVRE, Y. ANDROUSSI, and G. VANDERSCHAEVE
A TEM Investigation of the Dislocation Rosettes around Vickers Indentations in GaAs
(61.70; S7.12)
(a) Vol. **99**, No. 2, p. 405—412 (1987).

V. FILIPAVIČIUS, R. GAIDYS, V.-A. MATULAITIS, G. PETRAUSKAS, A. SAKALAS, and S. SAKALAUSKAS
Investigation of the Surface States on Heavily Doped GaAs by Kelvin Probe
(73.20; S7.12)
(a) Vol. **99**, No. 2, p. 543—547 (1987).

M. PROCOP, H. RAIDT, and B. SANDOW
Auger Electron and X-Ray Spectroscopic Analysis of Au–Ge Contacts to GaAs
(68.48; 73.40; 79.20; S1.3; S7.12)
(a) Vol. **99**, No. 2, p. 573—580 (1987).

I. THURZO and E. PINČÍK
Attempts to Find a Correlation between the GaAs Plasma Oxide Growth Conditions and the Charge DLTS Response of Interface States
(73.40; 71.55; S7.12)
(a) Vol. **99**, No. 2, p. K91—K95 (1987).

H. SIETHOFF, J. VÖLKL, D. GERTHSEN, and H. G. BRION
The Lower Yield Point of InP and GaAs
(62.20; S7.11; S7.12)
(a) Vol. **101**, No. 1, p. K13—K18 (1987).

C. MAJUMDAR, M. K. BOSE, A. B. MAITY, and A. N. CHAKRAVARTI
Effect of Size Quantization on Field Emission from Ultrathin Films of Degenerate Wide-Gap Semiconductors
(73.60; 79.70; S7.12)
(b) Vol. **141**, No. 2, p. 435—439 (1987).

M. ASCHE and O. G. SARBEI
Nonequilibrium Dynamics of Carriers and Phonons in GaAs after High Excitation by Short Time Laser Pulses
(71.38; 72.30; S7.12)
(b) Vol. **141**, No. 2, p. 487—491 (1987).

A. CHASSÉ and P. RENNERT
Theoretical Angle-Resolved X-Ray Photoelectron Spectra from GaAs (110)
(79.60; S7.12)
(b) Vol. **141**, No. 2, p. K147—K152 (1987).

H. YAMAMOTO
Resonant Tunneling through ABCBA-Type Double-Barrier Structures
(73.40; S5.11; S6; S7.12)
(b) Vol. **141**, No. 2, p. K179—K183 (1987).

L. GUOBA, G. ZUBAUSKAS, J. KAMENECKAS, and A. SAKALAS
Study of Multilayer Structures on GaAs by Rutherford Backscattering with Carbon Ions
(68.48; 79.20; S7.12)
(a) Vol. **101**, No. 2, p. 445—450 (1987).

G. KÜHNEL, W. SIEGEL, S. FELIX, and H. A. SCHNEIDER
On the Determination of the Conductivity and the Hall Coefficient of Semi-Insulating GaAs at Room Temperature
(72.20; S7.12)
(a) Vol. **101**, No. 2, p. 531—536 (1987).

W. HERGERT, P. RECK, L. PASEMANN, and J. SCHREIBER
Cathodoluminescence Measurements Using the Scanning Electron Microscope for the Determination of Semiconductor Parameters
(78.60; S7.12)
(a) Vol. **101**, No. 2, p. 611—618 (1987).

W. SIEGEL, H. KOI, G. KÜHNEL, and H. A. SCHNEIDER
Observation of the 78 meV Acceptor in p-Type GaAs Grown by the SSD Method
(71.55; 72.80; S7.12)
(a) Vol. **101**, No. 2, p. K147—K151 (1987).

W. Lorenz and T. Katterle
Chemical Processes on Semiconductor Electrode Interfaces and Distribution of Charge Transfer on the Orbital Energy Scale
(71.20 and 73.20; S7.12)
(b) Vol. **142**, No. 1, p. 149—163 (1987).

V. M. Jali, S. S. Kubakaddi, and B. G. Mulimani
Electronic Thermal Conductivity in Quasi-One-Dimensional Semiconductor Quantum Well Structures
(73.40; S7.12)
(b) Vol. **142**, No. 1, p. K25—K30 (1987).

L. Wendler and M. I. Kaganov
Helicon Waves in Semiconductor Superlattices under Quantum Hall Effect Conditions
(73.40; S7.12; S7.15)
(b) Vol. **142**, No. 1, p. K63—K68 (1987).

P. E. Simmonds and R. Sooryakumar
Stress Effects on Acceptor Bound Excitons in GaAs
(71.35 and 71.55; 78.55; S7.12)
(b) Vol. **142**, No. 2, p. K137—K141 (1987).

T. Figielski
Non-Stoichiometric Defects in GaAs Crystals
(61.70; S7.12)
(a) Vol. **102**, No. 2, p. 493—498 (1987).

Y. Nakada and T. Imura
Effects of Dopants and Deformation Temperature on Stacking Fault Energy in GaAs
(61.70; 62.20; 64.75; S7.12)
(a) Vol. **102**, No. 2, p. 625—632 (1987).

G. J. Papaioannou
Electron Radiation Effects on GaAs MESFETs
(61.80 and 85; 73.40; S7.12)
(a) Vol. **102**, No. 2, p. 843—848 (1987).

K. Kibickas, J. Parševiūnas, and Š. Vasiliauskas
Transient Hot Electron Phenomena in GaAs n^+–n–n^+ Structures at 300 K
(72.20 and 73.40; S7.12)
(a) Vol. **102**, No. 2, p. K99—K102 (1987).

M. K. Bose, C. Majumdar, A. B. Maity, and A. N. Chakravarti
On Field Emission from Ultrathin Films of Semiconductors with Effects of Image Forces
(73.60 and 79.70; S7.12)
(b) Vol. **143**, No. 1, p. 113—120 (1987).

W. Ulrici, J. Kreissl, A. Vasson, A. M. Vasson, and M. En-Naqadi
Photoinduced Recharging Phenomena and Vanadium-Related Centres in GaAs:V
(71.55; 76.30; 78.50; S7.12)
(b) Vol. **143**, No. 1, p. 195—206 (1987).

Y. Nakada and T. Imura
Thermal Stress and Stacking Fault Energy Estimated by TEM Observations of Grown-In Dislocations in Si-Doped GaAs Single Crystals
(61.70; S7.12)
(a) Vol. **103**, No. 1, p. 85—91 (1987).

R. Würschum and H.-E. Schaefer
Defects in Low-Temperature Electron-Irradiated GaAs Studied by Positrons
(61.70; 61.80; 78.70; S7.12)
(a) Vol. **103**, No. 1, p. 101—105 (1987).

H. Röppischer, R. Goldhahn, J. A. Bumai, and D. S. Domanevskii
Electroreflectance Depth Profiling and Hall Measurements of Thermally Converted GaAs Layers
(73.60 and 78.20; S7.12)
(a) Vol. **103**, No. 1, p. 161—166 (1987).

J. Pastrnak, F. Karel, and W. Ulrici
Two-Light-Source Extrinsic Photoconductivity of GaAs:V
(72.40; 71.55; S7.12)
(a) Vol. **103**, No. 1, p. 205—212 (1987).

A. I. Bobrysheva, S. S. Russu, and V. A. Zaloj
The Bihole in Quantum Well Structures
(71.25 and 73.40; S7.12; S7.15)
(b) Vol. **143**, No. 2, p. 549—555 (1987).

Xiaguang Wu, F. M. Peeters, and J. T. Devreese
Influence of Many-Body Effects on the Cyclotron Resonance Mass of Two-Dimensional Polarons with Application to GaAs–$Al_{1-x}Ga_xAs$ Heterostructures
(71.25 and 71.38; S7.12; S7.15)
(b) Vol. **143**, No. 2, p. 581—594 (1987).

B. Ulrici, R. Stedman, and W. Ulrici
Local Vibrational Mode Absorption of Hydrogen and Oxygen Centres in LEC-Grown GaP and GaAs
(63.20 and 78.30; 61.70; S7.11; S7.12)
(b) Vol. **143**, No. 2, p. K135—K139 (1987).

Y. Nakada and T. Imura
Influence of Dopants and Deformation Temperature on the Strength of GaAs
(62.20; 61.70; S7.12)
(a) Vol. **103**, No. 2, p. 435—442 (1987).

S7.13. Antimonides

K. P. Ghatak and M. Mondal
Theoretical Analysis of the Oscillatory Diffusivity–Mobility Ratio in Degenerate n-InSb in the Presence of a Quantizing Magnetic Field
(71.25; 72.20; S7.13)
(b) Vol. **139**, No. 1, p. 195—203 (1987).

T. Soma and H.-Matsuo Kagaya
Compression Effect on Elastic Coefficients of AlP, AlAs, and AlSb Compounds
(62.20; S7.11; S7.12; S7.13)
(b) Vol. **139**, No. 1, p. K1—K4 (1987).

V. I. Barbashov, V. I. Zaitsev, V. M. Mostovoi, and V. I. Alekseenko
Mobility of 60° Dislocations in Indium Antimonide on Reversing the Load
(61.70; S7.13)
(a) Vol. **100**, No. 2, p. K135—K138 (1987).

K. P. Maheshwari and G. Tarey
Stimulated Brillouin Scattering of EM Wave Propagating Obliquely in a Magnetoactive Solid State Plasma
(78.35 and 78.45; S7.13)
(b) Vol. **141**, No. 1, p. 333—341 (1987).

H.-U. Müller, F. Ludwig, and R. Herrmann
Cyclotron Resonance in the n-Inversion Layer of InSb Grain Boundaries
(72.20 and 76.40; S7.13)
(b) Vol. **142**, No. 1, p. K69—K71 (1987).

G. Gobsch, D. Schulze, and G. Paasch
Inversion Layers in InSb Bicrystals with a Charged Grain Boundary of Finite Thickness
(71.25 and 73.40; S7.13)
(b) Vol. **142**, No. 2, p. K119—K124 (1987).

M. de Dios Leyva, J. López Gondar, and J. Sabín del Valle
Nonparabolicity Parameter in the Conduction Band of GaAs–$Al_xGa_{1-x}As$ Quantum Wells
(71.25; S7.13; S7.15)
(b) Vol. **142**, No. 2, p. K151—K154 (1987).

B. Wolf and A. Zehe
Reaction Probability of Molecular Deuterium with a Disordered InSb (110) Surface
(68.45; S7.13)
(a) Vol. **104**, No. 1, p. K33—K35 (1987).

A. B. Maity, C. Majumdar, and A. N. Chakravarti
On Photoemission from Quasi-One-Dimensional Structures of Narrow-Gap Semiconductors Having Rectangular Cross-Sections
(73.30 and 79.60; S7.13)
(b) Vol. **144**, No. 2, p. K93—K98 (1987).

S7.15. Mixed crystals

G. MANZKE, K. HENNEBERGER, and V. MAY
Many-Exciton Theory for Multiple Quantum-Well Structures
(71.35; 78.20; S7.12; S7.15)
(b) Vol. **139**, No. 1, p. 233—239 (1987).

R. STEGMANN, W. N. RASBEGAEV, and A. N. PIKHTIN
Radiative Recombination in Zn–N Doped $GaAs_{1-x}P_x$:N Mixed Crystals
(78.55 and 78.60; 71.55; S7.15)
(a) Vol. **99**, No. 1, p. 289—294 (1987).

B. RHEINLÄNDER, J. KOVÁČ, P. NEVERMANN, and H. FIESELER
On the Determination of Quantum Yield in InGaAsP Double-Heterostructure LED
(73.40 and 85; S7.15)
(a) Vol. **99**, No. 1, p. 315—320 (1987).

H. HAEFNER, M. HEIDER, F. KOCH, R. MITDANK, G. OELGART, W. THIELEMANN, B. RHEINLÄNDER, and B. JACOBS
Properties of Red Light-Emitting (AlGa)As Single-Heterostructure Diodes. I. Structure Characterization
(73.40 and 85; 78.55; 78.60; S7.15)
(a) Vol. **99**, No. 2, p. 657—667 (1987).

J. LÓPEZ-GONDAR, H. RODRIGUEZ-COPPOLA, and R. PÉREZ-ALVAREZ
GaAs–AlGaAs Quantum Well of Compositional Graded Interface in the Envelope Function Approximation
(73.40; S7.12; S7.15)
(b) Vol. **140**, No. 1, p. 97—102 (1987).

M. I. ALIEV, KH. A. KHALILOV, and G. B. IBRAGIMOV
Free Carrier Absorption in $In_{1-x}Ga_xAs$
(78.30; S7.12; S7.15)
(b) Vol. **140**, No. 1, p. K83—K86 (1987).

B. RHEINLÄNDER, W. THIELEMANN, G. OELGART, H. HAEFNER, R. MITDANK, M. HEIDER, F. KOCH, and B. JACOBS
Properties of Red Light-Emitting (Al, Ga)As Single Heterostructure Diodes. II. Luminescence, Injection, and Photoeffect Efficiencies
(73.40 and 85; 78.55; 78.60; S7.15)
(a) Vol. **100**, No. 1, p. 369—377 (1987).

P. BASMAJI, M. GUITTARD, and P. GIBART
Enhancement of the Free Carriers Density in $Ga_{1-x}Al_xAs$ Grown by Metallorganic Vapor Phase Epitaxy under High Temperature Growth Conditions
(72.80; S7.15)
(a) Vol. **100**, No. 1, p. K41—K45 (1987).

B. KLOTH, M. V. LUPAL, A. N. PIKHTIN, C. E. RICHTER, R. RIES, R. STEGMANN, and M. TRAPP
Determination of the Nitrogen Concentration in VPE-$GaAs_{1-x}P_x$
(68.55 and 78.65; 64.75; S7.15)
(a) Vol. **100**, No. 2, p. 545—552 (1987).

L. Wendler and R. Pechstedt
Dynamical Screening, Collective Excitations, and Electron–Phonon Interaction in Heterostructures and Semiconductor Quantum Wells. Application to Double Heterostructures
(71.38 and 71.45; 68.30; 73.40; 77.30; 78.65; S7.12; S7.15)
(b) Vol. **141**, No. 1, p. 129—150 (1987).

L. Wendler and R. Haupt
Electron–Phonon Interaction in Semi-Infinite Semiconductor Superlattices
(71.38 and 71.45; 63.20; 68.30; 73.40; S7.1; S7.15)
(b) Vol. **141**, No. 2, p. 493—505 (1987).

L. Wendler and M. I. Kaganov
Helicon Waves in Semiconductor Superlattices under Quantum Hall Effect Conditions
(73.40; S7.12; S7.15)
(b) Vol. **142**, No. 1, p. K63—K68 (1987).

M. de Dios Leyva, J. López Gondar, and J. Sabín del Valle
Nonparabolicity Parameter in the Conduction Band of GaAs–$Al_xGa_{1-x}As$ Quantum Wells
(71.25; S7.13; S7.15)
(b) Vol. **142**, No. 2, p. K151—K154 (1987).

V. M. Jali, S. S. Kubakaddi, and B. G. Mulimani
On the Interpretation of Thermopower Measurements in GaAs–$Al_xGa_{1-x}As$ Heterostructures
(72.20 and 73.40; S7.15)
(b) Vol. **143**, No. 1, p. K25—K29 (1987).

R. Stegmann, B. Jacobs, M. Heider, M. Albani, and H. Kamleh
The Peculiarities of the Cathodoluminescence at Graded $Ga_{1-x}Al_xAs$ Heterojunctions
(73.40 and 78.60; S7.15)
(a) Vol. **103**, No. 1, p. 297—306 (1987).

L. Wendler and R. Haupt
Electron–Phonon Interaction in Semiconductor Superlattices
(71.38; 63.20; 68.30; 73.40; S7.1; S7.15)
(b) Vol. **143**, No. 2, p. 487—510 (1987).

A. I. Bobrysheva, S. S. Russu, and V. A. Zaloj
The Bihole in Quantum Well Structures
(71.25 and 73.40; S7.12; S7.15)
(b) Vol. **143**, No. 2, p. 549—555 (1987).

Xiaoguang Wu, F. M. Peeters, and J. T. Devreese
Influence of Many-Body Effects on the Cyclotron Resonance Mass of Two-Dimensional Polarons with Application to GaAs–$Al_{1-x}Ga_xAs$ Heterostructures
(71.25 and 71.38; S7.12; S7.15)
(b) Vol. **143**, No. 2, p. 581—594 (1987).

K. Nakashima, S. Nojima, Y. Kawamura, and H. Asahi
Deep Electron Trapping Centers in Si-Doped InAlAs Grown by Molecular Beam Epitaxy
(71.55; 72.40; S7.15)
(a) Vol. **103**, No. 2, p. 511—516 (1987).

R. Pérez-Alvarez, J. L. Parra-Santiago, and P. Pajón-Suárez
Cylindrical Quantum Well with Position Dependent Mass
(71.25; S7.12; S7.15)
(b) Vol. **144**, No. 2, p. 639—644 (1987).

S7.16. Higher compounds

M. Mondal and K. P. Ghatak
Theoretical Analysis of the Effective Electron Mass in n-Channel Inversion Layers on Ternary Chalcopyrite Semiconductors
(71.25; S7.16)
(b) Vol. **139**, No. 1, p. 185—193 (1987).

R. Gründler, T. Stöckert, H. Boudriot, K. Deus, and H. A. Schneider
Volume Plasmon Dispersion of Polycrystalline Films of the Ternary Semiconductors $ZnSnAs_2$ and $ZnSiAs_2$
(71.45; S7.16)
(b) Vol. **140**, No. 1, p. K19—K22 (1987).

D. Meimaris, J. Katris, D. Martakos, and G. P. Triberis
The Effect of Thermal Annealing on the Electrical Conductivity of Bulk Cd–Ge–As Glass
(72.80; S7.16)
(a) Vol. **100**, No. 2, p. K169—K172 (1987).

H. Boudriot, K. Deus, R. Gründler, H. A. Schneider, and T. Stöckert
Optical Properties of $ZnSiAs_2$ Single Crystalline Thin Films from Electron Energy Loss Measurements
(78.20 and 79.20; 71.25; S7.16)
(b) Vol. **143**, No. 2, p. K125—K130 (1987).

O. Uemura, Y. Takashima, K. Muraishi, and T. Satow
A Change in Local Order of Amorphous $CdGeAs_2$ on the Glass–Crystal Transition and the High Temperature Modification of Crystalline Phase
(61.50 and 61.60; S7.16)
(a) Vol. **103**, No. 2, p. K69—K74 (1987).

S8. Chalcogenides

(for transition metal compounds, see S1.63)

A. Masłowska
The Burstein Absorption Edge Structure in Lead Chalcogenides
(71.25; 71.38; 78.20; S8)
(b) Vol. **139**, No. 1, p. 161—171 (1987).

M. Inoue, M. Koyano, K. Fukushima, H. Negishi, and M. Sasaki
Pressure Dependence of Electrical Resistivity in Intercalation Compounds M_xTiS_2 (M = 3d Transition Metal)
(72.15 and 72.20; S8)
(b) Vol. **139**, No. 1, p. 273—280 (1987).

P. P. Petrov, V. Vassilev, and Z. Boncheva-Mladenova
Transport Properties of Monocrystalline Silver Telluride
(72.20; S8)
(a) Vol. **99**, No. 1, p. 237—242 (1987).

V. D. Vulchev and L. D. Borisova
Impurity States of Chromium in Lead Telluride
(71.55; 72.80; S8)
(a) Vol. **99**, No. 1, p. K53—K56 (1987).

P. Enders
Comment on "Electronic Properties of Tin Monochalcogenides (SnS, SnSe, SnTe)" by O. P. Singh and V. P. Gupta
(71.25; S8)
(b) Vol. **139**, No. 2, p. K129—K130 (1987).

R. M. Cândea, R. Turcu, G. Borodi, and I. Bratu
Effects of Thermal Annealing in Air on VE, COD, and CAD PbSe Films
(68.55 and 73.60; 78.65; S8)
(a) Vol. **100**, No. 1, p. 149—155 (1987).

B. Ray, P. A. Monsef Mirzai, and J. W. Brightwell
Fluorescence and ESR in CaS:Mn^{2+}
(71.55; 76.30; 78.55; S8)
(a) Vol. **100**, No. 1, p. 233—237 (1987).

H. Di Giulio, R. Rella, and A. Tepore
AC Electrical Conductivity in Amorphous Indium Selenide Thin Films
(73.60; S8)
(a) Vol. **100**, No. 1, p. K35—K39 (1987).

M. A. S. Sweet and J. Rennie
Calcium Sulphide as a Thermoluminescence γ-Ray and UV Dosimeter
(78.60; S8)
(a) Vol. **100**, No. 1, p. K67—K71 (1987).

V. Hönig and A. Thomas
Optical Constants of Ag_2Se in the Visible Spectral Range
(78.20; S8)
(a) Vol. **100**, No. 1, p. K81—K82 (1987).
Erratum
(a) Vol. **101**, No. 2, p. K183 (1987).

Kh. A. Mady, H. A. Mady, S. Y. Girgis, and H. R. Moustafa
Effect of Deposition Rate on Vacuum-Deposited PbS Films
(68.55; S8)
(a) Vol. **100**, No. 2, p. K107—K109 (1987).

G. Kluge
A New Interpretation of the Photodoping Effect in Amorphous As- and Ge-Chalcogenides
(66.30; 68.48; S8)
(a) Vol. **101**, No. 1, p. 105—114 (1987).

M. Wienecke and M. Schenk
In-Situ High-Temperature Hall and Conductivity Measurements for Investigations of the Stability Region of PbTe
(61.70; 66.30; 72.80; S8)
(a) Vol. **101**, No. 1, p. 115—122 (1987).

G. Kluge, R. Klabes, A. Thomas, R. Grötzschel, and P. Süptitz
High Energy Ion Beam Induced Ag Diffusion within Amorphous Ag:$GeSe_2$
(66.30; S8)
(a) Vol. **101**, No. 1, p. K9—K11 (1987).

A. Vohra, R. Verma, K. N. Lakshminarayan, and K. K. Srivastava
Dielectric Parameters for Planar Samples of Non-Conducting and Semiconducting Solid Powders at Microwave Frequencies
(77.20; S8)
(a) Vol. **101**, No. 1, p. K81—K85 (1987).

R. Mamy, A. Boufelja, and B. Carricaburu
Angle Resolved Photoemission and Electronic Band Structure of MoS_2
(71.25; 79.60; S8)
(b) Vol. **141**, No. 2, p. 467—473 (1987).

A. Cantarero, J. P. Martínez, A. Segura, and A. Chevy
Refractive Index of Bismuth Sulfide in the Infrared Region
(78.30; S8)
(a) Vol. **101**, No. 2, p. 603—609 (1987).

H. Berger, V. Brueckner, V. S. Dneprovskii, M. Graurock, and M. Lenzner
Investigation of Recombination Mechanisms in GaSe by Optoelectronic Switching
(72.40; S8)
(a) Vol. **101**, No. 2, p. K157—K160 (1987).

F. W. Boswell, A. Prodan, J. C. Bennetti, J. M. Corbett, and L. G. Hiltz
A Phase Transition and Domain Boundaries in the Modulation Structure of $TaTe_4$
(61.60; 61.70; S8)
(a) Vol. **102**, No. 1, p. 207—220 (1987).

J. P. Guesdon, B. Kobbi, C. Julien, and M. Balkanski
Electrical and Photovoltaic Properties of In_xSe_{1-x} Thin Films
(72.40 and 73.60; S8)
(a) Vol. **102**, No. 1, p. 327—335 (1987).

V. K. Lukyanyuk and Z. D. Kovalyuk
Sodium Intercalation into Indium and Gallium Selenides
(61.70; S8)
(a) Vol. **102**, No. 1, p. K1—K5 (1987).

Yu. P. Gnatenko and Yu. I. Zhirko
Exciton Absorption and Luminescence of Indium Selenide Crystals
(71.35 and 78.55; S8)
(b) Vol. **142**, No. 2, p. 595—604 (1987).

K. Takarabe, H. Kawamura, and K. Wakamura
Electronic Band Structure and Optical Anisotropy in InS. I. Experimental Study
(71.25 and 78.20; S8)
(b) Vol. **142**, No. 2, p. 605—615 (1987).

G. Dlubek, R. Krause, M. Mühlberg, and M. Schenk
Positron Study of Nonstoichiometric Vacancies in as-Grown p-PbTe Single Crystals
(61.70 and 78.70; S8)
(a) Vol. **102**, No. 2, p. K67—K70 (1987).

D. T. Tsiulyanu, E. P. Kolomeyko, and V. N. Stamov
Photoelectric Phenomena in Isotope Vitreous As_2S_3–Silicon Single-Crystal Heterojunctions
(72.40 and 73.40; S5.11; S8)
(a) Vol. **102**, No. 2, p. K103—K107 (1987).

J. E. Lewis
Optical Properties of the Alloy System SnTe–GeTe from Reflectance Measurements
(78.20; 71.25; 78.30; S8)
(b) Vol. **143**, No. 1, p. 307—315 (1987).

K. Takarabe
Raman Scattering Spectra of InS under Pressure
(78.30; S8)
(b) Vol. **143**, No. 1, p. K67—K70 (1987).

S8.1. II–VI compounds

K. P. GHATAK, N. CHATTERJEE, and M. MONDAL
Theoretical Analysis of the Carrier Effective Mass in Inversion Layers on II–VI Semiconductors
(71.25; S8.1)
(b) Vol. **139**, No. 1, p. K25—K30 (1987).

S. V. MELNICHUK, V. M. KRAMAR, and K. D. TOVSTYUK
The Influence of Covalent Bonding on the Amount of Spin–Orbit and Zeeman Interaction in $A^{II}B^{VI}$ Semiconductors Doped with 3d-Impurities
(71.70; S8.1)
(b) Vol. **141**, No. 2, p. K107—K111 (1987).

H.-MATSUO KAGAYA and T. SOMA
Thermal Properties of Tetrahedral Compounds
(65.40 and 65.70; S7.1; S8.1)
(b) Vol. **142**, No. 2, p. 411—416 (1987).

V. B. BIKBAEV, S. Č. KARPINSKAS, V. E. STOŠKUS, and J. J. VAITKUS
Forward Thermionic Emission Current of an Inhomogeneous Contact to Cadmium Chalcogenide
(73.30; S8.1)
(a) Vol. **102**, No. 2, p. 719—724 (1987).

YU. G. SEMENOV
Interaction of a Localized Spin Moment with Phonons Due to Exchange Scattering of Virtual Carriers in Semiconductors
(75.30; 75.80; 76.20; S8.1)
(b) Vol. **143**, No. 2, p. 717—726 (1987).

S8.11. Sulfides

F. Comas, C. Trallero Giner, and J. Tutor
Ohmic Mobility of Quasi-One-Dimensional Semiconducting Structures. Acoustic and Piezoelectric Phonon Scattering
(72.20 and 73.60; S7.12; S8.11)
(b) Vol. **139**, No. 2, p. 433—440 (1987).

H. Zimmermann and R. Boyn
Donor-Type Tm Centres in ZnS Crystals
(71.55; 78.55; S8.11)
(b) Vol. **139**, No. 2, p. 533—545 (1987).

J. Kreissl and D. Backs
EPR Investigations on Mn-Doped CVD-Grown ZnS Thin Film Layers
(76.30; S8.11)
(a) Vol. **99**, No. 2, p. K117—K120 (1987).

R. Boyn and H. Zimmermann
"Dissolution" of 4f–4f Transitions in Phonon Sidebands ($ZnS:Tm^{3+}$)
(71.55 and 78.55; 71.38; S8.11)
(b) Vol. **140**, No. 1, p. 163—172 (1987).

G. Czajkowski and P. Schillak
Interference Spectra of Semiconducting Slabs at Excitonic Resonances Studied by Stahl's Coherent Wave Approach to Band Gap Electrodynamics
(78.20; 71.35; S8.11; S8.12)
(b) Vol. **140**, No. 1, p. 273—278 (1987).

K. Löbe and R. Boyn
Dye-Laser Selective 4f–4f Excitation of Electroluminescent Sm^{3+} Centers in ZnS Films and Crystals
(78.55 and 78.60; S8.11)
(b) Vol. **140**, No. 1, p. K91—K95 (1987).

K. Matsuura, S. Kishida, and I. Tsurumi
Annealing of the F^+ EPR Centers and the 2.3 and 2.9 eV Optical Absorption Bands in Zinc Sulfide Crystals
(61.70; 76.30; 78.50; S8.11)
(b) Vol. **140**, No. 2, p. 347—354 (1987).

R. K. Singh and S. Singh
Study of Elastic Properties and Their Pressure Dependence of Zincblende Structure Semiconductors
(62.20; S8.11; S8.12; S8.13)
(b) Vol. **140**, No. 2, p. 407—413 (1987).

J. Nolly, K. K. Abdullah, and K. P. Vijayakumar
Ellipsometric Study of Spray Pyrolysed CdS Thin Films
(68.55 and 78.65; S8.11)
(a) Vol. **101**, No. 1, p. K35—K38 (1987).

Cz. Koepke
Cavityless Optical Bistability in Thin Sprayed CdS Films
(78.20; S8.11)
(b) Vol. **141**, No. 2, p. K139—K142 (1987).

M. I. Strashnikova, V. Ya. Reznichenko, and V. V. Cherny
Size Effect in the Refractive Index Dispersion of Thin CdS Crystals
(78.20; S8.11)
(b) Vol. **141**, No. 2, p. K153—K156 (1987).

P. Singh and B. Baishya
Some Studies on Thin Film Transistors (TFTs) with Er_2O_3 as Gate Insulator
(73.40 and 85; S8.11; S8.12)
(a) Vol. **101**, No. 2, p. K173—K177 (1987).

K. Matsuura, S. Kishida, I. Tsurumi, and M. Kitagawa
Electron-Irradiation Effects on the F^+ Bands and the Near-Infrared Emissions in ZnS Crystals
(78.50 and 78.55; 61.70; S8.11)
(b) Vol. **142**, No. 1, p. K79—K83 (1987).

K. Matsuura, S. Kishida, K. Yoshida, and I. Tsurumi
Optical Bleaching of the F^+ Optical Absorption Bands in ZnS Crystals
(61.70 and 78.50; S8.11)
(b) Vol. **142**, No. 2, p. 617—627 (1987).

K. Matsuura, S. Kishida, Y. Fukata, and I. Tsurumi
Correlation between the 1.1 and 1.45 eV Emission Bands and the F^+ Optical Absorption Bands in ZnS Crystals
(78.50 and 78.55; 61.70; S8.11)
(b) Vol. **143**, No. 1, p. 275—280 (1987).

M. Kretzschmar, F. Henneberger, H. Rossmann, and I. Haddad
Dynamical Effects in Increasing Absorption Optical Bistability of Thermal Origin
(78.20; S8.11)
(b) Vol. **143**, No. 1, p. K71—K76 (1987).

M. Hanke, D. Hennig, and A. Kaschte
A Tight-Binding Description of Isovalent Impurity Clusters. Application to Te Impurities in II–VI Compound Semiconductors
(71.55; S8.11; S8.12; S8.13)
(b) Vol. **143**, No. 2, p. 655—661 (1987).

S8.12. Selenides

K. Singh and J. P. Rai
Electrosynthesis and Photoelectroactivity of Polycrystalline p-Zinc Selenide
(72.40; 61.60; S8.12)
(a) Vol. **99**, No. 1, p. 257—265 (1987).

V. A. Kasiyan, D. D. Nedeoglo, A. V. Simashkevich, and I. N. Timchenko
Magnetoresistivity of ZnSe in the Region of Hopping Conduction over Localized States in the Coulomb Gap
(72.20; 71.55; S8.12)
(b) Vol. **139**, No. 2, p. 559—564 (1987).

J. Sekulski
On the Possibility of Application of the Derivative Lightoutput–Voltage Characteristic for the Analysis of Electron–Phonon Interaction in Electroluminescent ZnSe MOS Structures
(73.40 and 78.60; S8.12)
(a) Vol. **99**, No. 2, p. K139—K141 (1987).

G. Czajkowski and P. Schillak
Interference Spectra of Semiconducting Slabs at Excitonic Resonances Studied by Stahl's Coherent Wave Approach to Band Gap Electrodynamics
(78.20; 71.35; S8.11; S8.12)
(b) Vol. **140**, No. 1, p. 273—278 (1987).

D. Boulanger and R. Parrot
Spin–Lattice Coupling Coefficients of Mn^{2+} In ZnSe
(76.30; S8.12)
(b) Vol. **140**, No. 1, p. K79—K82 (1987).

R. K. Singh and S. Singh
Study of Elastic Properties and Their Pressure Dependence of Zincblende Structure Semiconductors
(62.20; S8.11; S8.12; S8.13)
(b) Vol. **140**, No. 2, p. 407—413 (1987).

N. A. Dovgaleva, A. F. Plotnikov, V. A. Tolokonnikov, V. E. Shubin, V. I. Botsan, and V. A. Kasian
Photoelectric Properties of the Ni–ZnSe–Ge Structure
(72.40 and 73.40; S8.12)
(a) Vol. **100**, No. 2, p. 527—533 (1987).

I. E. Türe, A. W. Brinkman, G. J. Russell, and J. Woods
The Effects of Oxygen in Au–CdSe Schottky Diodes
(85; 68.20; 73.30; S8.12)
(a) Vol. **100**, No. 2, p. 681—691 (1987).

H. Pradella and U. W. Pohl
Luminescence of the Cr^{+} Centre in ZnSe
(78.55; S8.12)
(b) Vol. **141**, No. 1, p. K143—K146 (1987).

P. Singh and B. Baishya
Some Studies on Thin Film Transistors (TFTs) with Er_2O_3 as Gate Insulator
(73.40 and 85; S8.11; S8.12)
(a) Vol. **101**, No. 2, p. K173—K177 (1987).

M. Hanke, D. Hennig, and A. Kaschte
A Tight-Binding Description of Isovalent Impurity Clusters. Application to Te Impurities in II–VI Compound Semiconductors
(71.55; S8.11; S8.12; S8.13)
(b) Vol. **143**, No. 2, p. 655—661 (1987).

H. L. Oczkowski
Kinetic Model with an Intra-Pair Metastable State for Thermally Stimulated Processes
(71.55; 78.60; S8.12)
(a) Vol. **103**, No. 2, p. 499—509 (1987).

S. Kishida, K. Matsuura, H. Nagase, and I. Tsurumi
Annealing Characteristics and Temperature Dependence of the Photosensitive 1.9 eV Absorption Band in Zn-Treated ZnSe Crystals
(78.50; 61.70; S8.12)
(a) Vol. **103**, No. 2, p. 613—618 (1987).

M. S. Brodin, A. O. Gushcha, V. N. Khotiaintsev, S. G. Shevel, L. V. Taranenko, and V. V. Tishchenko
Nonlinearities in Excitonic Luminescence of Direct-Gap Semiconductors under Very Low Optical Excitation Levels
(78.55; S8.12; S9)
(b) Vol. **144**, No. 2, p. 863—873 (1987).

S8.13. Tellurides

P. Schäfer and H. Niebsch
Growth of (100) and (111) CdTe Films on (100) GaAs by Hot-Wall Epitaxy
(68.55; S8.13)
(a) Vol. **99**, No. 1, p. K41—K44 (1987).

P. H. E. Meijer, P. Pecheur, and G. Toussaint
Electronic Structure of the Cd Vacancy in CdTe
(71.55; 61.70; 71.25; S8.13)
(b) Vol. **140**, No. 1, p. 155—162 (1987).

M. Samimi, B. Biglari, M. Hage-Ali, J. M. Koebel, and P. Siffert
About the Origin of the 0.15 to 0.20 eV Defect Level in Cadmium Telluride
(71.55; S8.13)
(a) Vol. **100**, No. 1, p. 251—258 (1987).

K. Pastor and R. Triboulet
A Model for Persistent Cyclotron Resonance in Pure n-CdTe
(72.40 and 76.40; S8.13)
(a) Vol. **100**, No. 1, p. K51—K56 (1987).

S8.15. Mixed crystals

V. I. Ivanov-Omskii, N. N. Berchenko, and A. I. Elizarov
Transport Anomalies in $Cd_xHg_{1-x}Te$ **(Review Article)**
(72.80; 73.25; S8.15)
(a) Vol. **103**, No. 1, p. 11—28 (1987).

S. F. Chehab and J. C. Woolley
Energy Gap and Valence Band Structure of $Cd_{1-x}Mn_xTe_{1-y}Se_y$ Alloys
(71.25; 78.20; S8.15)
(b) Vol. **139**, No. 1, p. 213—222 (1987).

M. Fathallah, M. Ben Said, and R. Bennaceur
Determination of the Low Absorption Coefficients of Thin $Cd_{1-x}Zn_xS$ Films by Photothermal Deflection Spectroscopy
(78.65; S8.15)
(a) Vol. **99**, No. 2, p. 521—526 (1987).

G. Ch. Christakudis, S. K. Plachkova, E. S. Avilov, L. E. Shelimova, and N. Kh. Abrikosov
Electrical Resistivity and Scattering Mechanisms of $(GeTe)_{1-x}((Ag_2Te)_{1-y}(Sb_2Te_3)_y)_x$ Solid Solutions ($y = 0.6$)
(72.20; S8.15)
(a) Vol. **99**, No. 2, p. 593—601 (1987).

S. M. Wasim and R. Tovar B.
On the Thermal Conductivity of $CuInSe_{2x}Se_{2(1-x)}$
(66.70; S8.15)
(a) Vol. **99**, No. 2, p. K69—K73 (1987).

J. Mašek and B. Velický
Mn 3d States in Photoelectron Spectra from $Cd_{1-x}Mn_xTe$
(71.25; 79.60; S8.15)
(b) Vol. **140**, No. 1, p. 135—140 (1987).

Z. Dziuba
Analysis of the Electrical Conduction in CdHgTe Crystals
(72.20; S8.15)
(b) Vol. **140**, No. 1, p. 213—223 (1987).

M. Předota, L. Beneš, and J. Horák
On the Incorporation of Germanium Atoms into the Bi_2Te_3 Crystal Lattice
(61.60; S8.15)
(a) Vol. **100**, No. 2, p. 401—404 (1987).

N. E. Korsunskaya, I. V. Markevich, and N. R. Stratieva
Photoelectrical and Optical Properties of $Zn_xCd_{1-x}S$ Layers
(72.40 and 73.60; 78.65; S8.15)
(a) Vol. **100**, No. 2, p. 521—525 (1987).

Z. Kučera
Dispersion of the Refractive Index of $Hg_{1-x}Cd_xTe$
(78.20; S8.15)
(a) Vol. **100**, No. 2, p. 659—665 (1987).

S8.16. Higher compounds

P. J. S. FOOT, T. KATZ, S. N. PATEL, B. A. NEVETT, A. R. PIERCY, and A. A. BALCHIN
The Structures and Conduction Mechanisms of Lithium-Intercalated and Lithium-Substituted Nickel Phosphorus Trisulphide ($NiPS_3$), and the Use of the Material as a Secondary Battery Electrode **(Review Article)**
(72.20; 61.60; S8.16)
(a) Vol. **100**, No. 1, p. 11—29 (1987).

L. I. HAWORTH, I. S. AL-SAFFAR, and R. D. TOMLINSON
Electrical Characteristics of Melt Grown $CuInTe_2$ Single Crystals
(72.80; 71.55; S8.16)
(a) Vol. **99**, No. 2, p. 603—610 (1987).

L. K. SAMANTA
On Some Properties of I_2–IV–VI_3 Compounds
(78.20; S8.16)
(a) Vol. **100**, No. 1, p. K93—K97 (1987).

M. A. ALDZHANOV, N. G. GUSEINOV, and Z. N. MAMEDOV
Heat Capacity of $TlGaSe_2$
(65.40; S8.16)
(a) Vol. **100**, No. 2, p. K145—K148 (1987).

F. M. KAMARLY and A. YU. YANGIROV
Magnetic Susceptibility of the $(SnTe)_{1-x}$–$(EuTe)_x$ System
(75.20; 72.80; S8.16)
(a) Vol. **101**, No. 1, p. K61—K63 (1987).

S. M. PATEL and V. G. KAPALE
$AgGaSe_2$/Al Thin Film Schottky Barrier Diodes
(73.30 and 85; S8.16)
(a) Vol. **101**, No. 1, p. K87—K92 (1987).

S. N. TOUBEKTSIS, J. A. KALOMIROS, L. V. GIOBLIAKIS, K. M. PARASKEVOPOULOS, A. N. ANAGNOSTOPOULOS, and E. K. POLYCHRONIADIS
On the Growth and Characterization of the System $TlBiSe_2$–$TlBiTe_2$
(61.60; 72.80; 78.20; S8.16)
(a) Vol. **101**, No. 2, p. 355—360 (1987).

I. V. BODNAR, B. V. KORZUN, and A. P. CHERNYAKOVA
Microhardness of the $A^IB^{III}C_2^{VI}$ Ternary Semiconductors and Their Solid Solutions
(62.20; S8.16)
(a) Vol. **101**, No. 2, p. 409—419 (1987).

N. S. ORLOVA and I. V. BODNAR
The Anisotropy of the Thermal Expansion in the $AgInC_2^{VI}$ Chalcopyrite-Type Compounds from the X-Ray Diffraction Data
(65.70; S8.16)
(a) Vol. **101**, No. 2, p. 421—425 (1987).

H. NEUMANN, G. KÜHN, and W. MÖLLER
Heat Capacity and Lattice Anharmonicity in Cu–III–VI_2 Chalcopyrite Compounds
(65.40; S8.16)
(b) Vol. **144**, No. 2, p. 565—573 (1987).

G. A. MEDVEDKIN, YU. V. RUD, and M. A. TAIROV
Optical Band Edge Absorption in $CuInSe_2$
(78.20; S8.16)
(b) Vol. **144**, No. 2, p. 809—816 (1987).

P. GARD, C. SOURISSEAU, and O. GOROCHOV
Effect of Metal Substitution in ZrS_3 and TiS_3 Compounds. Electronic, Raman, and Resonance Raman Study of $Zr_{1-x}Ti_xS_3$ Ternary Phases, $0 < x < 1$
(78.30; 63.20; S8.16)
(b) Vol. **144**, No. 2, p. 885—901 (1987).

K. D. BRONSEMA and J. MAHY
One-, Two-, and Three-Dimensional Order in and Four-Dimensional Description of $K_xV_5S_8$ $(0.5 \leqq x \leqq 0.7)$
(61.60; S8.16)
(a) Vol. **104**, No. 2, p. 603—618 (1987).

B. VENGATESAN, N. KANNIAH, and P. RAMASAMY
CVT Growth, Microhardness, and Electrical Characterization of $ZnGa_2S_4$ Single Crystals
(62.20 and 72.80; S8.16)
(a) Vol. **104**, No. 2, p. K93—K96 (1987).

S9. Halides

W. Y. Zeng, J. F. Higgs, and A. Anderson
Lattice Dynamics of Tin Tetrabromide
(63.20; S9)
(b) Vol. **139**, No. 1, p. 85—96 (1987).

J. Lakshmana Rao
Optical Absorption Spectrum of Cr^{3+} Ions Doped in NH_4Cl Single Crystals
(71.55; 78.50; S9)
(b) Vol. **139**, No. 1, p. 241—244 (1987).

N. V. Unnikrishnan, H. S. Bhatti, and R. D. Singh
Two-Photon Photoconductivity in CdI_2 under Double Beam Laser Excitation
(72.40; S9)
(b) Vol. **139**, No. 1, p. K41—K45 (1987).

M. Saleem, M. Kadri, and R. C. Agrawal
Transport Properties of a Cationic Superionic Conductor
(66.30; S9)
(a) Vol. **99**, No. 1, p. 91—95 (1987).

M. Kumari, K. Kumari, and N. Dass
On the Melting Law at High Pressures
(64.70; S2; S9)
(a) Vol. **99**, No. 1, p. K23—K26 (1987).

D. Neogy and T. Purohit
The Behavior of Active Centers in a Laser Host. A Crystal Field Investigation of Sm^{3+} in LaF_3 Single Crystal
(71.70; 65.40; 75.30; S9)
(b) Vol. **139**, No. 1, p. 519—525 (1987).

J. C. Krupa, C. Khan Malek, P. Delamoye, B. Moine, and C. Pedrini
Observation of an Incommensurate Structure by Optical Spectroscopy with a 5f Ion as an Internal Probe. U^{4+}–$ThCl_4$ System
(78.50 and 78.55; 71.70; S9)
(b) Vol. **140**, No. 1, p. 289—300 (1987).

P. Négrier, N. B. Chanh, C. Courseille, C. Hauw, A. Méresse, and M. Couzi
The Imma ↔ Pnma Phase Transition of $NH_3(CH_2)_5NH_3CdCl_4$ Studied by X-Ray Diffraction
(64.60; 61.10; S9)
(a) Vol. **100**, No. 2, p. 473—483 (1987).

HUANG MING
High-Order Perturbation Formula for the EPR Trigonal Symmetric Parameter D of $d^2(d^8)$ Ions and the Determination of the $NiCl_2$ Crystal Structure
(61.60 and 76.30; S9)
(b) Vol. **141**, No. 1, p. K1—K6 (1987).

P. LABORDE, G. VILLENEUVE, and J. M. REAU
NMR Study of the Mobility of Fluor Ions in $Pb_{1-x}Bi_xO_xF_{2-x}$ Solid Solutions
(66.30; 76.60; S9)
(b) Vol. **141**, No. 2, p. 379—390 (1987).

M. H. KUOK
Raman Spectra of Cesium Trichlorostannate(II)
(78.30; 63.20; S9)
(b) Vol. **141**, No. 2, p. K167—K171 (1987).

M. DAS and A. K. PAL
EPR Studies on Phase Transitions in Copper Tetramethylammonium Bromide
(64.60; 76.30; S9)
(a) Vol. **101**, No. 2, p. 403—408 (1987).

M. SALEEM, S. B. HALIM, and R. C. ARAWAL
Electrical Conduction in the Superionic Material $(Ag_2Se)_{0.452}(AgI)_{0.258}(HgI_2)_{0.290}$
(66.30; S9)
(a) Vol. **102**, No. 1, p. K29—K32 (1987).

N. SHUKLA and G. C. TRIGUNAYAT
Study of Structural Defects in Melt-Grown Cadmium Iodide Crystals
(61.50; 61.70; S9)
(a) Vol. **103**, No. 2, p. 337—347 (1987).

SHANG BO, HUANG MING, and MIN-GUANG ZHAO
The Relationship between the Pressure and Zero-Field Splitting of $ZnSiF_6 \cdot 6\,H_2O{:}Ni^{2+}$ and the Determination of Its Equations of State
(71.70; S9)
(b) Vol. **144**, No. 1, p. K35—K38 (1987).

M. S. BRODIN, A. O. GUSHCHA, V. N. KHOTIAINTSEV, S. G. SHEVEL, L. V. TARANENKO, and V. V. TISHCHENKO
Nonlinearities in Excitonic Luminescence of Direct-Gap Semiconductors under Very Low Optical Excitation Levels
(78.55; S8.12; S9)
(b) Vol. **144**, No. 2, p. 863—873 (1987).

T. SHIOZAWA and T. KOBAYASHI
High Resolution Electron Microscopic Study of Photolytic Silver in Silver Bromide
(61.80; 61.16; S1.3; S9)
(a) Vol. **104**, No. 2, p. 649—659 (1987).

S9.11. Alkali halides

L. BOSI and M. NIMIS
Lifetime Studies on Excited $(F_2^+)^*$ and M Centers in NaF Doped with Magnesium
(61.70; 78.50; S9.11)
(b) Vol. **140**, No. 2, p. 355—360 (1987).

B. G. DICK
Microscopic Models for Substitutional OH^- and CN^- Defects in Alkali-Halide Crystals
(61.70; S9.11)
(b) Vol. **141**, No. 1, p. 61—70 (1987).

R. K. SINGH and N. V. K. PRABHAKAR
On the Phase Transition in Sodium Halides
(64.70; S9.11)
(b) Vol. **141**, No. 1, p. K29—K33 (1987).

M. MATOS
UHF MC INDO Calculations of the Interstitial Hydrogen in Alkali Halides
(61.70; 71.55; S9.11)
(b) Vol. **141**, No. 2, p. 399—408 (1987).

F. J. LOHMEIER
Photostimulated Reorientations of PO_2^- Centers in Alkali Halides
(71.55; 78.55; S9.11)
(b) Vol. **141**, No. 2, p. 517—521 (1987).

E. CARRILLO H., A. CARRILLO H., and J. RUBIO O.
A Simple Geometric Model for Predicting Coherence between Matrix and Precipitated Phases in Alkali Halides Doped with Divalent Impurities
(61.50; 61.70; S9.11)
(a) Vol. **101**, No. 2, p. 315—327 (1987).

L. BOSI
A Simple Theory for Forecasting the Melting Temperature of Alkali Halides and of Alkaline-Earth Oxides
(64.70; 77.20; S9.11; S10.1)
(a) Vol. **101**, No. 2, p. K111—K114 (1987).

J. OTTO
Investigation of the Phonon Sidebands of SH^- and SeH^- Centres in Cesium Halides and Potassium Bromide
(63.20 and 78.30; S9.11)
(b) Vol. **142**, No. 1, p. 105—115 (1987).

M. KITAMURA, S. MURAMATSU, and C. SUGIURA
Multiple Scattering Study for Cl K XANES of KCl, $SrCl_2$, and CsCl Crystals
(71.10 and 78.70; S9.11; S9.12)
(b) Vol. **142**, No. 1, p. 191—202 (1987).

T. OHGAKU and N. TAKEUCHI
The Blaha Effect of Alkali Halide Crystals
(62.20; S9.11)
(a) Vol. **102**, No. 1, p. 293—299 (1987).

C. N. Afonso, I. M. Gomez, and C. Ortiz
A New Luminescent Band Induced in NaF by Ion Implantation
(78.55; 61.70; S9.11)
(b) Vol. **143**, No. 2, p. 777—782 (1987).

M. Nadgornyi and H. P. Strunk
Dislocation Structure of Plastically Deformed NaCl Polycrystals
(61.70; 62.20; S9.11)
(a) Vol. **104**, No. 1, p. 193—202 (1987).

L. S. Fomenko, H.-J. Kaufmann, S. V. Lubenets, V. D. Natsik, V. V. Pustovalov, and S. E. Shumilin
Anomalies in Plastic Response of High-Purity LiF Single Crystals at 0.55 to 4.2 K
(62.20; S9.11)
(a) Vol. **104**, No. 1, p. K7—K11 (1987).

D. Cardenas-García, E. R. López-Tellez, and C. Ruíz-Mejía
Calculations of the Polarization Energy of a Complex in Alkali Halide Crystals from Zero to Fourth Order
(61.70; S9.11)
(b) Vol. **144**, No. 2, p. 477—486 (1987).

L. Bosi and M. Nimis
Effect of Thermal Expansion on Absorption Band Shift of the F-Center in Alkali Halides
(61.70; 78.50; S9.11)
(b) Vol. **144**, No. 2, p. 501—505 (1987).

M. Kumar, D. Hans, A. K. Sharma, and J. Shanker
Structural Phase Transition of Alkali Halides Using a Modified Interaction Model
(64.60; 61.50; S9.11)
(b) Vol. **144**, No. 2, p. 543—548 (1987).

L. N. Kantorovich
Multiple Theory of the Polarization of Solids by Point Defects. IV. Electronic and Spatial Structure of Single Electronic and Hole Centers and Their Pairs in LiF, KCl Crystals
(71.55; 77.30; S9.11)
(b) Vol. **144**, No. 2, p. 719—726 (1987).

J. L. Jansons, V. J. Krumins, Z. A. Rachko, and J. A. Valbis
Luminescence Due to Radiative Transition between Valence Band and Upper Core Band in Ionic Crystals (Crossluminescence)
(78.60; 71.20; S9.11; S9.12)
(b) Vol. **144**, No. 2, p. 835—844 (1987).

H.-B. Abel and F. Fischer
Additional Results on SERS from Ag Colloids in Alkali Halide Crystals
(78.30; 61.70; S9.11)
(b) Vol. **144**, No. 2, p. 875—884 (1987).

E. Radzhabov
Vacuum Ultraviolet Absorption and Dissociation of H_a^- and OH^- Centers in NaF
(78.50; 61.70; 71.55; S9.11)
(b) Vol. **144**, No. 2, p. K151—K155 (1987).

S9.12. Alkaline earth halides

R. Ubgade and P. R. Sarode
Study of Strontium Compounds and Minerals by X-Ray Absorption Spectroscopy. I. K-Edge Shifts
(78.70; S3; S9.12; S10; S11.1)
(a) Vol. **99**, No. 1, p. 295—301 (1987).

R. K. Singh and C. N. Rao
Analysis of Cohesive and Anharmonic Properties of Alkaline Earth Fluoride Crystals
(62.20; S9.12)
(b) Vol. **139**, No. 2, p. 393—398 (1987).

S. Kubota, N. Kanai, and J. Ruan(Gen)
Fine Structure of the Fast Luminescence Component from BaF_2 Crystal Excited by Fast Electrons
(78.60; S9.12)
(b) Vol. **139**, No. 2, p. 635—639 (1987).

B. M. Bolotov, D. A. Aksenov, A. I. Polyakov, and S. P. Senshin
NMR Line Shape Function Analysis in Solids
(76.60; S4; S9.11; S9.12)
(b) Vol. **139**, No. 2, p. K137—K140 (1987).

A. Davituliani, M. Kiknadze, R. Mirianashvili, and T. Sanadze
RFDS Studies of Trigonal Er^{3+} Centres in BaF_2 Single Crystals
(76.70; S9.12)
(b) Vol. **140**, No. 1, p. K69—K73 (1987).

A. Hamaïda and J. Margerie
Measurement of Oscillator Strengths of "F_{2A}" Centres in $MeF_2:A^+$ (Me = Ca, Sr; A^+ = Li^+, Na^+, K^+)
(61.70; 76.30; 78.50; S9.12)
(b) Vol. **141**, No. 2, p. 391—398 (1987).

U. Subramanian and M. L. Mukherjee
Dielectric Studies of CaF_2:Nd Single Crystals from RT to 350 °C
(77.20 and 77.40; S9.12)
(a) Vol. **101**, No. 2, p. 577—581 (1987).

M. Kitamura, S. Muramatsu, and C. Sugiura
Multiple Scattering Study for Cl K XANES of KCl, $SrCl_2$, and CsCl Crystals
(71.10 and 78.70; S9.11; S9.12)
(b) Vol. **142**, No. 1, p. 191—202 (1987).

Sp. Dedoussis, G. D. Vassiliades, A. Liolios, C. Touramanis, M. Chardalas, C. A. Elefteriades, and St. Charalambous
Positron Annihilation Studies in Ionic Crystals
(78.70; 61.80; S9.11; S9.12)
(a) Vol. **102**, No. 2, p. 577—581 (1987).

V. K. Jain and M. S. Jahan
Thermal Activation Energies and Peak Temperatures in Thermoluminescence of LiF(Mg, Ti) and CaF_2:Mn at Low Temperatures
(71.55 and 78.60; S9.11; S9.12)
(a) Vol. **102**, No. 2, p. 725—731 (1987).

Jia-Jun Chen and Min-Guang Zhao
High-Order Perturbation Formulae of the Spin-Hamiltonian Parameters for the Ground State $^3A_2(F)$ of d^8 Ions in Rhombic Symmetry
(71.70; S9.12)
(b) Vol. **143**, No. 2, p. 647—653 (1987).

J. L. Rao, R. M. Krishna, and S. V. J. Lakshman
Nearest-Neighbour Point Ion Approximation Calculations of Energy Levels for Mn^{2+} Ions Doped in MgF_2 Single Crystals
(71.55; 78.50; S9.12)
(b) Vol. **143**, No. 2, p. K99—K101 (1987).

Y. Ravi Sekhar and H. Bill
^2S Ground State Ions in Ionic Crystals: ESR Results
(76.30; 61.70; S9.12; S10)
(b) Vol. **143**, No. 2, p. K111—K114 (1987).

J. L. Jansons, V. J. Krumins, Z. A. Rachko, and J. A. Valbis
Luminescence Due to Radiative Transitions between Valence Band and Upper Core Band in Ionic Crystals (Crossluminescence)
(78.60; 71.20; S9.11; S9.12)
(b) Vol. **144**, No. 2, p. 835—844 (1987).

N. Guskos
An Interpretation of the Electric Field Effect on Cubic Gd^{3+} Centers in BaF_2 Single Crystals
(76.30; 71.70; S9.12)
(b) Vol. **144**, No. 2, p. K139—K142 (1987).

R. V. Gopala Rao and U. Bandyopadhyay
Partial Structure Factors and Associated Correlation Functions of Molten Strontium Chloride
(61.25; S9.12)
(a) Vol. **104**, No. 2, p. 575—585 (1987).

S9.15. Mixed crystals

A. Memon and D. B. Tanner
Phonon Combination Bands in the Far-Infrared Spectrum of $K_{0.5}Rb_{0.5}I$ Mixed Crystals
(63.20 and 78.30; S9.15)
(b) Vol. **139**, No. 1, p. K81—K85 (1987).

L. Sirdeshmukh, G. Sathaiah, and P. Devi
Dielectric Properties of the KCl–KBr Mixed Crystal System
(77.20 and 77.40; S9.15)
(a) Vol. **99**, No. 2, p. 631—639 (1987).

M. T. Sprackling and M. E. Bahrololoom
The Role of Iodide in the Strain-Ageing of Silver Iodobromide
(61.70; 62.20; S9.15)
(a) Vol. **100**, No. 1, p. 105—108 (1987).

J. Shanker, S. C. Sharma, and M. Kumar
Analysis of the Structural Phase Transition in Mixed Alkali Halide Crystals
(64.60; S9.15)
(b) Vol. **141**, No. 2, p. 409—414 (1987).

B. R. K. Gupta
Lattice Vibrations in Mixed Alkali Halide Crystals
(63.20; S9.15)
(b) Vol. **142**, No. 1, p. 77—81 (1987).

J. Aaltonen and R. Lindström
Effect of Uniaxial Stress on the Interdiffusion in the KCl–KBr System
(66.30; S9.15)
(a) Vol. **102**, No. 2, p. K71—K75 (1987).

H. Ketata, M. H. Ben Ghozlen, A. Daoud, and I. Pabst
Low-Temperature Structure (T = 235 K) and Structural Phase Transition in $[(C_2H_5)_4N]_2SnCl_6$
(61.60; 64.70; S9.15)
(a) Vol. **103**, No. 1, p. 73—78 (1987).

Yu. N. Orlov, V. I. Aleshin, G. E. Arkhangelskii, V. E. Bozhevolnov, Yu. V. Voronov, L. N. Ivanov, V. V. Karelin, and E. I. Saunin
Concentration Peculiarities of a Set of Impurity Centres in Fluorite Crystals Activated by Gadolinium
(61.70; 76.30; 78.60; S9.15)
(b) Vol. **144**, No. 2, p. K77—K80 (1987).

S9.16. Higher compounds

M. K. TENG, J. F. VITTORI, M. MASSOT, and S. ZIOLKIEWICZ
Dielectric and Ferroelectric Properties in $As_xSb_{1-x}SI$ Mixed Crystals
(77.20 and 77.80; S9.16)
(a) Vol. **99**, No. 2, p. 641—647 (1987).

E. SCHULTHEISS, A. SCHARMANN, and D. SCHWABE
Energy Levels of Triply Ionized Praseodymium in $BiLiF_4$ Crystals
(71.70 and 78.40; 63.20; S9.16)
(b) Vol. **140**, No. 1, p. 173—189 (1987).

K. VENU, V. S. S. SASTRY, and J. RAMAKRISHNA
Molecular Dynamics in $[(CH_3)_4N]_2CdI_4$. A Proton Magnetic Relaxation Study
(76.60; 63.90; S9.16)
(b) Vol. **140**, No. 1, p. 251—259 (1987).

C. SUGIURA, M. KITAMURA, and S. MURAMATSU
Ruthenium L_{II} and Rhodium $L_{II,III}$ X-Ray Absorption-Edge Structures
(78.70; S9.16)
(b) Vol. **140**, No. 2, p. 631—636 (1987).

A. WOJAKOWSKA and E. KUNDYS
High Electrical Conductivity of a Solid Phase in the SnI_2–AgI System
(72.80; 64.70; S9.16)
(a) Vol. **101**, No. 2, p. K153—K155 (1987).

S. ABDI, Y. MLIK, and M. H. BEN GHOZLEN
Group-Theoretical Analysis of the Symmetry Changes in R_2MX_4 Compounds
(61.50; 64.70; S9.16)
(b) Vol. **142**, No. 1, p. 27—34 (1987).

M. V. R. REDDY, J. S. KUMAR, K. N. REDDY, and U. V. S. RAO
Dielectric Properties of $NaYF_4$ Thin Films
(77.55; S9.16)
(a) Vol. **102**, No. 1, p. 321—326 (1987).

SHUEN WEI-LI and ZHAO MIN-GUANG
Spin–Orbit and Crystal-Field Matrices of $d^{4,6}$ Configuration and Application to $RbFeCl_3$
(71.70; S9.16)
(b) Vol. **142**, No. 2, p. 501—515 (1987).

R. EWERTOWSKI and W. ŚWIĄTKOWSKI
Annihilation of Positron Trapped at the Ag/Na_3AlF_6 Interface
(78.70; 68.48; S9.16)
(a) Vol. **102**, No. 2, p. 555—557 (1987).

S10. Oxides

A. L. Ivanovskii, D. L. Novikov, and V. A. Gubanov
Electronic Properties of Carbides, Nitrides, and Oxides of Subgroup VIa Transition Metals. Theory and Experiment **(Review Article)**
(71.25; 73.20; 78.70; 79.60; S1.61; S1.62; S10)
(b) Vol. **141**, No. 1, p. 9—33 (1987).

A. G. Pelmenshchikov, G. M. Zhidomirov, S. Beran, and J. Tiňo
Molecular Approach in the Study of the Processes of Formation of the Lattice of Oxides. Characteristic Features of Coordination of AlO_3 Structural Elements
(61.50; S10)
(a) Vol. **99**, No. 1, p. 57—64 (1987).

K. Zakrzewska, T. Pisarkiewicz, and A. Czapla
Scattering of Charge Carriers in Reactively Sputtered $CdIn_2O_4$ Thin Films
(73.60; 78.65; S10; S11.2)
(a) Vol. **99**, No. 1, p. 141—149 (1987).

R. Ubgade and P. R. Sarode
Study of Strontium Compounds and Minerals by X-Ray Absorption Spectroscopy. I. K-Edge Shifts
(78.70; S3; S9.12; S10; S11.1)
(a) Vol. **99**, No. 1, p. 295—301 (1987).

M. Klanjšek Gunde, B. Orel, and V. Moiseenko
Infrared Reflection Spectra of and Polariton Dispersion in Ferroelectric $Pb_5Ge_3O_{11}$
(63.20 and 78.30; S10)
(b) Vol. **139**, No. 2, p. K151—K155 (1987).

M. H. Islam and C. A. Hogarth
Some Optical and Structural Studies on SiO/SnO_2 Thin Films
(78.65; 68.55; 71.25; S10)
(a) Vol. **99**, No. 2, p. 503—512 (1987).

Y. Fukuda, A. Tomita, and N. Takeuchi
Thermoluminescence and Thermally Stimulated Exoelectron Emission of Sintered CaB_4O_7 Doped with Pb, Eu, or Dy
(78.60 and 79.75; S10)
(a) Vol. **99**, No. 2, p. K135—K138 (1987).

C. S. Sunandana
Structure and Cu^{++} EPR of γ'-$Pb_3V_2O_8$
(61.60; 64.70; 76.30; S10)
(a) Vol. **101**, No. 1, p. 37—43 (1987).

Hong Sun and Junxin Fang
Polariton Theory of Infrared Absorption Edges in SiO_2, GeO_2, and GeO_2-Doped SiO_2 Glasses
(78.30; 71.36; S10)
(b) Vol. **143**, No. 1, p. 289—297 (1987).

V. Dvořák
On the Structural Phase Transition in La_2CuO_4
(63.20 and 64.70; S10)
(b) Vol. **143**, No. 1, p. K15—K18 (1987).

H. W. Zandbergen, G. van Tendeloo, T. Okabe, and S. Amelinckx
Electron Diffraction and Electron Microscopy of the Superconducting Compounds $Ba_2YCu_3O_{7-\delta}$
(61.60; 61.14; 61.16; S10)
(a) Vol. **103**, No. 1, p. 45—72 (1987).

D. Y. Parpia, M. Safa, B. K. Tanner, and G. F. Clark
Synchrotron X-Ray Topography Observation of the Morin Transition in Highly Perfect Crystals of Haematite
(75.60; 61.10; S10)
(a) Vol. **103**, No. 1, p. 263—271 (1987).

G. van Tendeloo and S. Amelinckx
Direct Observation of the Order–Disorder Transformation on the Oxygen Sublattice of $Ba_2YCu_3O_{7-\delta}$
(61.50; S10)
(a) Vol. **103**, No. 1, p. K1—K6 (1987).

O. Smirčková, D. Sýkorová, J. Dominec, K. Jurek, and L. Smrčka
On the Superconductivity in the Y–Ba–Cu–O System
(74.70; S10)
(a) Vol. **103**, No. 1, p. K33—K36 (1987).

K. Bartkowski, R. Horyń, A. J. Zaleski, Z. Bukowski, M. Horobiowski, C. Marucha, J. Rafalowicz, K. Rogacki, A. Stępień-Damm, C. Sułkowski, E. Trojnar, and J. Klamut
Thermal and Electrical Conductivity of $La_{1.8}Sr_{0.2}CuO_4$ Samples
(74.70; S10)
(a) Vol. **103**, No. 1, p. K37—K39 (1987).

Y. Ravi Sekhar and H. Bill
^2S Ground State Ions in Ionic Crystals: ESR Results
(76.30; 61.70; S9.12; S10)
(b) Vol. **143**, No. 2, p. K111—K114 (1987).

G. Blasse
Luminescence from off-Centre Pb^{2+} Ions
(78.55; S10)
(b) Vol. **143**, No. 2, K131—K 133 (1987).

S10.1. Simple oxides

R. Wang and D. Liu
Foil Thickness Determination by CBED in the Kinematical Approximation
(61.14; S1.2; S4; S10.1)
(a) Vol. **99**, No. 1, p. 47—55 (1987).

V. I. Sokolov and N. A. Fedorovich
Mechanical Stresses on the Si–SiO_2 Interface
(68.25; S5.11; S10.1)
(a) Vol. **99**, No. 1, p. 151—158 (1987).

J. Fontcuberta
Lattice Dynamics Study of Polycrystalline ^{57}CoO by Mössbauer Spectroscopy
(63.20; 76.80; S10.1)
(b) Vol. **139**, No. 2, p. 379—386 (1987).

C. Lee
Configurations of Small Si/O Clusters
(61.70; S10.1)
(b) Vol. **139**, No. 2, p. K93—K97 (1987).

J. Bąk-Misiuk, J. Gronkowski, J. Härtwig, and W. Wierzchowski
X-Ray Spherical-Wave Diffraction in the Bragg Case for Bent Si Crystals
(61.10; 61.55; S5.11; S10.1)
(a) Vol. **99**, No. 2, p. 345—351 (1987).

E. G. Salman, G. V. Perov, and V. N. Vertoprakhov
Ionic Current Study in Thermally Grown Silicon Dioxide Films on Polycrystalline Silicon
(73.40 and 77.30; 77.55; S10.1)
(a) Vol. **99**, No. 2, p. 537—542 (1987).

A. J. Abu El-Haija and A. H. Madjid
DC Electrical Conductivity of Ultrathin Ag–SiO Superlattices
(73.40; S1.3; S10.1)
(b) Vol. **140**, No. 1, p. 113—120 (1987).

B. Balland, A. Glachant, A. Ronda, and J. C. Dupuy
High-Temperature Thermal Nitridation and Low-Temperature Electron-Beam-Enhanced Nitridation of SiO_2 Thin Films
(68.48 and 73.40; S5.11; S10.1)
(a) Vol. **100**, No. 1, p. 187—198 (1987).

D. Köhler and E. Kubalek
Spatially Resolved Cathodoluminescence Investigations on Zinc Oxide Ceramics
(78.60; 61.70; S10.1)
(a) Vol. **100**, No. 1, p. 337—342 (1987).

Ďurček, R. Fettig, and J. Rosenzweig
The Charge Injection Effect on the Acousto-Electric Properties of $Si-SiO_2$ Structures
(72.50 and 73.40; S5.11; S10.1)
(a) Vol. **101**, No. 1, p. K33—K34 (1987).

R. Lacroix
Spin Hamiltonian of Cr^{3+} in a Strained Ruby Crystal
(71.70; 62.20; S10.1)
(b) Vol. **141**, No. 2, p. 567—573 (1987).

V. Kumar
Bond Susceptibility and Ionicity in $A^{III}B^{V}C_4^{VI}$ Compounds
(61.50 and 77.20; S10; S10.1)
(b) Vol. **141**, No. 2, p. K157—K162 (1987).

R. I. Joseph and M. E. Thomas
Differential Reflection Spectroscopy: A Direct Determination of Long Wavelength Optical Mode Frequencies in Polar Crystals
(78.20; S10.1)
(b) Vol. **141**, No. 2, p. K163—K166 (1987).

J. Kirchhof, P. Kleinert, W. Radloff, and E. Below
Diffusion Processes in Lightguide Materials. The Diffusion of OH in Silica Glass at High Temperatures
(66.30; 64.75; S10.1)
(a) Vol. **101**, No. 2, p. 391—401 (1987).

I. M. Silvestrova, Y. V. Pisarevskii, P. A. Senyushenkov, A. I. Krupny, R. Voszka, I. Földvári, and J. Janszky
Temperature Dependence of Elastic Properties of Paratellurite
(62.20; 62.65; 65.70; S10.1)
(a) Vol. **101**, No. 2, p. 437—444 (1987).

A. E. Rakhshani and J. Varghese
Optical Absorption Coefficient and Thickness Measurement of Electrodeposited Films of Cu_2O
(78.65; S10.1)
(a) Vol. **101**, No. 2, p. 479—486 (1987).

S. A. Bhosale, G. S. Nadkarni, and S. Radhakrishnan
Negative Capacitance in Thin Film $Al-V_2O_5-Al$ Devices
(73.40 and 85; S10.1)
(a) Vol. **101**, No. 2, p. 639—646 (1987).

L. Bosi
A Simple Theory for Forecasting the Melting Temperature of Alkali Halides and of Alkaline-Earth Oxides
(64.70; 77.20; S9.11; S10.1)
(a) Vol. **101**, No. 2, p. K111—K114 (1987).

A. Remón, J. A. García, M. A. Ochando, J. Llopis, and J. Piqueras
Photoluminescence from Heat-Treated MgO:Ni Crystals
(78.55; S10.1)
(a) Vol. **101**, No. 2, p. K167—K171 (1987).

N. SASI, C. BALASUBRAMANIAN, and S. K. NARAYANDASS
Structure, Dielectric, and AC Conduction Studies on Germanium Dioxide Thin Films
(68.55 and 73.60; 77.50; S10.1)
(a) Vol. **103**, No. 2, p. 475—480 (1987).

T. STAPIŃSKI, E. JAPA, and J. ZUKROWSKI
^{119}Sn Mössbauer Investigation of Cadmium–Tin–Oxide Thin Films
(68.55 and 76.80; S10.1)
(a) Vol. **103**, No. 2, p. K93—K98 (1987).

H. J. BLYTHE, J. RIVAS, and J. IÑIGUEZ
An Investigation of the Magnetic After-Effect in Magnetite at High Temperatures
(75.60; S10.1)
(a) Vol. **103**, No. 2, p. K137—K140 (1987).

V. P. GRIBKOVSKI, A. L. GURSKII, G. A. PASHKEVICH, and G. P. YABLONSKII
Crystallographic Orientation of Electric Discharges in TeO_2 and $LiNbO_3$ Monocrystals
(77.50; 63.20; S10.1; S11.1)
(a) Vol. **103**, No. 2, p. K153—K156 (1987).

XIAO-PENG HE and MAO-LU DU
An EPR Study of the Structure of the Cr^{3+}-Vacancy Centre in MgO
(76.30; S10.1)
(b) Vol. **144**, No. 1, p. K51—K56 (1987).

K. TENNAKONE, S. W. M. S. WICKRAMANAYAKE, P. SAMARSEKARA, and C. A. N. FERNANDO
Doping of Semiconductor Particles with Salts
(71.55 and 72.80; S10.1)
(a) Vol. **104**, No. 1, p. K57—K60 (1987).

N. AMIN, S. ARAJS, and E. MATIJEVIC
Magnetic Properties of Submicronic α-Fe_2O_3 Particles of Uniform Size Distribution at 300 K
(75.50; S10.1)
(a) Vol. **104**, No. 1, p. K65—K68 (1987).

A. A. BEREZIN
Localized Electron Levels in Isotopically Ordered Crystals (Isotopic Quantum Wells)
(71.50; S10.1)
(b) Vol. **144**, No. 2, p. 727—732 (1987).

M. OKADA, T. KAWAKUBO, T. SEIYAMA, and M. NAKAGAWA
Enhancement of 3d-Electron Transitions in Neutron-Irradiated MgO:Mn^{2+} Crystals
(78.50; 71.70; S10.1)
(b) Vol. **144**, No. 2, p. 903—909 (1987).

B. W. LICZNERSKI, H. PAGNIA, and N. SOTNIK
Conductivity Model of Cermets Treated as Critical Percolative Systems. Application to Thick Film Resistors
(73.40; S10.1)
(a) Vol. **104**, No. 2, p. 891—902 (1987).

S10.15. Mixed crystals

AL. NICULA and E. CULEA
Determination of the Distribution Widths of the Spin Hamiltonian Parameters of V^{4+} Ions from Oxide Glasses
(76.30; 61.40; S10.15)
(b) Vol. **142**, No. 1, p. 265—270 (1987).

R. HERRMANN, N. KUBICKI, TH. SCHURIG, H. DWELK, U. PREPPERNAU, A. KRAPF, H. KRÜGER, H.-U. MÜLLER, L. ROTHKIRCH, W. KRAAK, W. BRAUNE, N. PRUSS, G. NACHTWEI, F. LUDWIG, and E. KEMNITZ
Preparation and Properties of High T_c Superconducting Y–Ba–Cu–O Systems
(74.70; S10.15)
(b) Vol. **142**, No. 1, p. K53—K56 (1987).

C. F. VAN BRUGGEN
Monovalent-Copper (I; $3d^{10}$, $S = 0$) Induced Oxygen-2p Valence Band Holes with Concomitant Attractive Interaction inside Low-Dimensional Oxygen Networks
(71.25 and 74.70; 75.20; S1.4; S10.15)
(b) Vol. **142**, No. 2, p. 537—550 (1987).

J. K. SRIVASTAVA, S. MURALIDHARA RAO, R. NAGARAJAN, and R. VIJAYARAGHAVAN
Mössbauer Effect, Ionic Conductivity, and Ionic Thermoconductivity Studies of Ionic Spin Fluctuations in the Magnetically Ordered Fe^{3+} : $NiTiO_3$ System
(76.80; 66.30; S10.15)
(a) Vol. **102**, No. 2, p. 759—768 (1987).

G. CAMPET, SUN ZHI WEN, C. PUPRICHITKUN, J. P. MANAUD, and J. CLAVERIE
Physical Properties of $(n^+)SrTiO_3$ Films
(73.60; S10.15)
(a) Vol. **103**, No. 1, p. 175—184 (1987).

R. PROVENCHER, M. E. BRETT, and M. G. TOWNSEND
Superconductivity and Materials Preparation in $YBa_2Cu_3O_{6+x}$
(74.70; S10.15)
(a) Vol. **103**, No. 2, p. K107—K111 (1987).

YUHUAN XU, ZHONGRONG LI, CHAORUI LI, HONG WANG, WU LI, and XIANDUAN LIN
La and Li Doped High-T_c Superconductor of Y–Ba–Cu–O Ceramics
(74.70; S10.15)
(a) Vol. **103**, No. 2, p. K113—K117 (1987).

J. KAMARÁD, Z. ARNOLD, and E. POLLERT
Pressure Dependence of the Critical Temperature of Superconductivity in $YBa_2Cu_3O_{7-x}$
(74.10; S10.15)
(b) Vol. **144**, No. 1, p. K39—K43 (1987).

G. KRABBES, R. MÜLLER, M. RITSCHEL, H. VINZELBERG, E. WOLF, B. A. AMINOV, NGUEN MIN TCHU, M. V. SUDAKOVA, and YA. G. PONOMAREV
Josephson Effects in Superconducting Y–Ba–Cu–O at Liquid Nitrogen Temperatures
(75.50; 74.70; S10.15)
(a) Vol. **104**, No. 1, p. K61—K64 (1987).

S11. Other higher compounds

DU MAO-LU and ZHAO MIN-GUANG
A Theoretical Explanation for the Pressure Shift of Optical Spectra in Chromium Potassium Alum
(71.70 and 78.40; S11)
(b) Vol. **139**, No. 1, p. K87—K91 (1987).

YU WAN-LUN and ZHAO MIN-GUANG
Zero-Field Splitting and the d–d Transitions of Mn^{2+} on Ca(II) Sites in $Ca_5(PO_4)_3F$
(71.70; S11)
(b) Vol. **140**, No. 1, p. 203—212 (1987).

J. TRITT-GOC, N. PISLEWSKI, and R. GOC
NMR Study of Flip Motion of the Water Molecules in SNP
(76.60; S11)
(a) Vol. **100**, No. 1, p. K57—K59 (1987).

J. MÁLEK and M. VANĚČEK
The Absorption Edge in the n-Type $Ge_{20}Bi_{15}S_{65}$ Glass
(78.40; S11)
(a) Vol. **100**, No. 1, p. K99—K102 (1987).

M. DUBIEL and D. EHRT
Nuclear Magnetic Resonance Investigations of the Structure of Fluorophosphate Glasses
(61.40; 76.60; S11)
(a) Vol. **100**, No. 2, p. 415—422 (1987).

M. JURCZYK
Crystallographic and Magnetic Characteristics of $Ce_3Co_{20}B$ Compound
(61.60 and 75.50; S11)
(a) Vol. **100**, No. 2, p. K173—K176 (1987).

K. BIELECKI, Z. KRUCZYŃSKI, and S. I. FĂRCAŞ
On New Results of Cr^{3+} EPR in Potassium Aluminium Alum Crystals
(76.30; S11)
(b) Vol. **141**, No. 1, p. K67—K70 (1987).

A. K. GEORGIEV, M. A. MISHEVA, and G. KH. TOUMBEV
TDPAC Study of $HfO(NO_3)_2 \cdot 2\,H_2O$
(65.90 and 76.80; S11)
(a) Vol. **101**, No. 1, p. K19—K21 (1987).

M. ONODA and H. NAGASAWA
Phase Transition and Bipolaron Formation in Quasi-One-Dimensional Conductor β-$Ag_xV_2O_5$
(64.60 and 71.38; 76.30; 76.70; S11)
(b) Vol. **141**, No. 2, p. 507—516 (1987).

I. Śledzińska
Electric Field Gradient in Ferrous Oxalate Dihydrate
(76.80; S11)
(b) Vol. **143**, No. 1, p. K63—K66 (1987).

O. A. Golikova and M. M. Kazanin
Modified Aluminium Dodecaboride
(72.80 and 78.30; S11)
(a) Vol. **103**, No. 1, p. K41—K43 (1987).

K. Tennakone, A. H. Jayatissa, C. A. N. Fernando, S. Wickramanayake, S. Punchihewa, L. K. Weerasena, and W. D. R. Premasiri
Semiconducting and Photoelectrochemical Properties of n- and p-Type β-CuCNS
(73.60 and 78.65; 68.55; 72.40; S11)
(a) Vol. **103**, No. 2, p. 491—497 (1987).

T. S. N. Murthy and M. Salagram
Electronic Spectra of Radiation-Induced CrO_4^{3-} in 3 $CdSO_4 \cdot 8\, H_2O$ Crystals
(61.80 and 71.70; 78.50; S11)
(b) Vol. **144**, No. 2, p. 519—527 (1987).

S11.1. Salts of oxy-acids

A. K. Pradhan, R. N. P. Choudhary, and B. M. Wanklyn
Raman and Infrared Spectra of $YAsO_4$
(78.30; S11.1)
(b) Vol. **139**, No. 1, p. 337—345 (1987).

L. Cianchi, A. Cuccoli, M. Mancini, and G. Spina
A Comment on the Role of the Mixed Spin Relaxation Process in $ZnCO_3:Fe^{2+}$
(75.30; S11.1)
(b) Vol. **139**, No. 1, p. K53—K56 (1987).

S. Ahmed, A. M. Shamah, R. Kamel, and Y. Badr
Structural Changes of $(NH_4)_2SO_4$ Crystals
(64.70; 77.80; S11.1)
(a) Vol. **99**, No. 1, p. 131—140 (1987).

W. Kaczmarek and F. Gervais
Infrared Dispersion of β-Gadolinium Molybdate at Room Temperature
(78.30; S11.1)
(a) Vol. **99**, No. 1, p. 279—287 (1987).

J. K. Rath and S. Radhakrishna
EPR Studies of $MoO^3{}_4{}^-$ Centres in KDP Single Crystals
(77.80; 76.30; S11.1)
(a) Vol. **100**, No. 1, p. 331—336 (1987).

H. Hess
The Fatiguing of the Photoluminescence in Zinc Silicate Doped with Mn and Ga
(78.55; 72.40; S11.1)
(a) Vol. **100**, No. 1, p. 343—354 (1987).

A. M. Balagurov, A. I. Beskrovnii, B. N. Savenko, B. V. Merinov, M. Dlouhá, S. Vratislav, and Z. Jirák
The Room Temperature Structure of Deuterated $CsHSO_4$ and $CsHSeO_4$
(61.50; S11.1)
(a) Vol. **100**, No. 1, p. K3—K7 (1987).

A. García-Cabañes, C. Prieto, and L. Arizmendi
X-Ray Damage in $LiTaO_3$
(61.80; 76.30; S11.1)
(a) Vol. **100**, No. 1, p. K9—K12 (1987).

P. Górski and W. Kucharczyk
On the Measurement of the Quadratic Electrooptic Effect in Noncentrosymmetric Crystals
(78.20; S11.1)
(a) Vol. **100**, No. 1, p. K73—K76 (1987).

Chen Nandon and Gu Xuerong
On the Optical Anomaly of $LiNbO_3$ above Room Temperature
(78.20; S11.1)
(a) Vol. **100**, No. 1, p. K77—K80 (1987).

F. Lisse, J. Köhler, H. Pufahl, and D. Schmid
Sensitized Luminescence of Pure and Doped $NaNO_2$ Single Crystals. I. Stationary Spectra
(78.55; 71.35; 71.55; S11.1)
(b) Vol. **140**, No. 2, p. 605—619 (1987).

E. Cazzanelli, A. Fontana, F. Rocca, and R. Frech
Low-Temperature Behaviour of the Lithium Modes in Monoclinic Li_2SO_4
(78.30; 63.20; S11.1)
(b) Vol. **140**, No. 2, p. 621—630 (1987).

Y. A. Badr, S. B. El-Giziri, and T. M. Hammad
On the Effect of Doping on the Mechanism of Phase Transition of Some Ionic Inorganic Compounds
(64.70 and 77.80; 72.80; 77.70; S11.1)
(a) Vol. **100**, No. 2, p. 467—472 (1987).

Z. Jirák, M. Dlouhá, S. Vratislav, A. M. Balagurov, A. I. Beskrovnyi, V. I. Gordelii, I. D. Datt, and L. A. Shuvalov
A Neutron Diffraction Study of the Superionic Phase in $CsHSO_4$
(61.60 and 64.70; S11.1)
(a) Vol. **100**, No. 2, p. K117—K122 (1987).

B. T. Deshmukh, S. V. Bodade, and S. V. Moharil
Thermoluminescence and Reflectance Spectra of $K_2Ca_2(SO_4)_3$ and $K_2Mg_2(SO_4)_3$
(78.60; 64.70; S11.1)
(a) Vol. **102**, No. 1, p. 381—388 (1987).

C. Jäger, S. Barth, A. Feltz, and G. Scheler
Studies of the Structure of Mixed Crystals of the System $Na_{1+2x}Mg_xZr_{2-x}(PO_4)_3$ by ^{31}P-MAS-NMR
(76.60; 61.60; S11.1)
(a) Vol. **102**, No. 2, p. 791—802 (1987).

S. Ahmed, A. M. Shamah, R. Kamel, and Y. Badr
Phase Transition in $[(NH_4)_{1-x}K_x]_2SO_4$ Crystals
(64.70; S11.1)
(a) Vol. **102**, No. 2, p. K81—K85 (1987).

Wen-cheng Zheng
Theoretical Calculations of EPR Parameters and Absorption Spectra in Chromium-Doped α-$LiIO_3$ Crystals and Determination of Local Compressibilities
(71.70; 62.20; 76.30; 78.50; S11.1)
(b) Vol. **143**, No. 1, p. 217—223 (1987).

A. M. Balagurov, F. Prokert, and B. N. Savenko
Phase Transition Effects in the Incommensurate Modulated Single Crystal of $Sr_{0.7}Ba_{0.3}Nb_2O_6$
(64.70; 61.60; S11.1)
(a) Vol. **103**, No. 1, p. 131—144 (1987).

A. E. Ovechkin, V. D. Ryzhikov, G. Tamulaitis, and A. Žukauskas
Luminescence of $ZnWO_4$ and $CdWO_4$ Crystals
(78.55; S11.1)
(a) Vol. **103**, No. 1, p. 285—290 (1987).

P. Górski and W. Kucharczyk
The Quadratic Electrooptic Effect in KDP and ADP Crystals
(78.20; S11.1)
(a) Vol. **103** No. 1, p. K65—K67 (1987).

A. Förster, S. Kapphan, and M. Wöhlecke
Overtone Spectroscopy of the OH and OD Stretch Modes in $LiNbO_3$
(78.30 and 78.50; S11.1)
(b) Vol. **143**, No. 2, p. 755—764 (1987).

O. Thiemann and O. F. Schirmer
Microwave Electric Field Induced ESR of Ni^{2+} in $LiNbO_3$
(76.30; S11.1)
(b) Vol. **143**, No. 2, p. K115—K118 (1987).

Bhuwan Chandra, S. S. Shinde, A. R. Lakshmanan, and R. C. Bhatt
Effect of Doping with Monovalent Cation Impurities on the Thermoluminescence Response of $CaSO_4$: Dy Phosphor
(78.60; S11.1)
(a) Vol. **103**, No. 2, p. 599—605 (1987).

S11.2. Spinels, garnets, ferrites

K. Zakrzewska, T. Pisarkiewicz, and A. Czapla
Scattering of Charge Carriers in Reactively Sputtered $CdIn_2O_4$ Thin Films
(73.60; 78.65; S10; S11.2)
(a) Vol. **99**, No. 1, p. 141—149 (1987).

K. M. Reddy, A. S. Jacob, B. J. Reddy, and Y. P. Reddy
Optical Absorption Spectra of Cu^{2+} in Brochantite
(78.30 and 78.40; S11.2)
(b) Vol. **139**, No. 2, p. K145—K150 (1987).

J. K. Srivastava, K. Muraleedharan, and R. Vijayaraghavan
Entropic Ferrimagnetism of $FeNiCrO_4$ System
(75.50; 76.80; S11.2)
(b) Vol. **140**, No. 1, p. K47—K50 (1987).

J. Čermák, L. Kalivoda, and M. Nevřiva
LPE Films of Magnetic Garnets with Bi Substitution
(68.55 and 75.70; S11.2)
(a) Vol. **100**, No. 1, p. 213—219 (1987).

J. K. Srivastava, K. Muraleedharan, and R. Vijayaraghavan
Anomalous Mössbauer Lineshapes in Mixed Spinel Ferrite $Ga_{0.6}Fe_{0.4}NiCrO_4$
(75.30 and 76.80; S11.2)
(b) Vol. **140**, No. 2, p. K137—K139 (1987).

M. Wurlitzer
Manifestations of Induced Magnetic Anisotropy in Zr-Doped Yttrium Iron Garnet
(75.50; S11.2)
(a) Vol. **100**, No. 2, p. 607—620 (1987).

P. V. Reddy and M. Salagram
The Far-Infrared Spectra of Mixed Manganese–Magnesium Ferrites
(78.30; S11.2)
(a) Vol. **100**, No. 2, p. 639—643 (1987).

Satbir Singh, R. B. Tripathi, and B. K. Das
Disaccomodation in GeO_2/SnO_2-Doped Ni–Zn Ferrites
(75.30; S11.2)
(a) Vol. **100**, No. 2, p. K185—K188 (1987).

J. S. Diez Viñuela and C. Otero Areán
Distribution of Copper Ions among Octahedral and Tetrahedral Sites in $Cu_xMg_{1-x}Al_2O_4$ Spinels
(61.60; S11.2)
(a) Vol. **101**, No. 1, p. 57—61 (1987).

R. Hergt, H. Pfeiffer, P. Görnert, M. Wendt, B. Keszei, and J. Vandlik
Kinetic Segregation of Lead Impurities in Garnet LPE Films
(64.75; 68.45; 68.55; S11.2)
(a) Vol. **104**, No. 2, p. 769—776 (1987).

A. I. Eaṭah, A. A. Ghani, M. F. El-Shahat, and E. El-Faramawy
Temperature Dependence of Thermoelectric Power and Electrical Resistivity in $Cu_xCo_{1-x}Fe_2O_4$ Ferrites
(72.20; S11.2)
(a) Vol. **104**, No. 2, p. 793—797 (1987).

K. Hisatake, I. Matsubara, K. Maeda, T. Fujihara, N. Ichinose, I. Sasaki, and T. Nakano
Photoinduced Effects on the Permeability in YIG Single Crystals Grown by the Floating Zone Method
(75.50; S11.2)
(a) Vol. **104**, No. 2, p. 815—824 (1987).

J. Wiehl and S. Kemmler-Sack
Cathodo- und Photoluminescence in the Eu^{3+} Activated Garnets of Type $Y_{3-x}Ln_xTe_2Li_3O_{12}$ (Ln = Lu, La) and $Gd_{3-x}La_xTe_2Li_3P_{12}$
(78.55 and 78.60; 78.30; S11.2)
(a) Vol. **104**, No. 2, p. 863—872 (1987).

E. Z. Katsnelson
Photogalvanomagnetic Properties of Manganese Ferrite
(72.40; S11.2)
(a) Vol. **104**, No. 2, p. K127—K132 (1987).

S12. Organic compounds

V. G. Podoprigora, I. A. Remizov, V. F. Shabanov, and A. N. Botvich
Surface Phonons in Molecular Crystals
(68.30; S12)
(b) Vol. **139**, No. 1, p. 109—115 (1987).

D. Sarkar and T. N. Misra
Compensation Effect in the Electrical Conduction Process in Some Nucleic Acid Base Complexes with Acridine Orange Dye
(72.80; S12)
(a) Vol. **99**, No. 1, p. 251—256 (1987).

J. A. Giacometti and R. Coelho
Further Speculations on the Natural Decay of Corona Charged Insulating Films
(73.25; S12)
(a) Vol. **99**, No. 1, p. K45—K48 (1987).

F. Engelke, D. Michel, W. Windsch, and J. Petersson
^{2}D NMR Study near the Ferroelectric Phase Transition of TSCC
(76.60 and 77.80; S12)
(b) Vol. **139**, No. 2, p. 609—617 (1987).

F. Povolo and R. E. Bolmaro
Average Elastic Constants and Tensor Invariants
(62.20; S1.2; S3; S11.1; S12)
(a) Vol. **99**, No. 2, p. 423—436 (1987).

M. Zor and C. A. Hogarth
High-Field Electrical Conduction in Thin Films of Polyethylene
(73.60; S12)
(a) Vol. **99**, No. 2, p. 513—519 (1987).

N. K. Sanyal, R. A. Yadav, and S. R. Shukla
Propagation of Ultrasonic Waves in a Smectic Liquid Crystal. Cholesteryl Erucate
(62.60; 64.70; S12)
(a) Vol. **100**, No. 1, p. 109—114 (1987).

R. Singh, J. M. Keller, and S. C. Datt
Thermally Stimulated Current and Charge Decay Studies in Polypropylene
(77.55; S12)
(a) Vol. **100**, No. 1, p. 143—147 (1987).

K. Biedrzycki and B. Kosturek
Thermally Stimulated Electron Emission from Triglycine Sulphate Crystals with Well-Defined Initial Domain Structure
(77.80 and 79.40; S12)
(a) Vol. **100**, No. 1, p. 327—330 (1987).

H. Kresse and P. Rabenstein
Geometrical Aspects and Dipole–Dipole Interaction in a Liquid Crystalline Mixture
(61.25 and 77.20; S12)
(a) Vol. **100**, No. 1, p. K83—K85 (1987).

S. Nešpůrek, I. Koropecký, M. Šorm, A. Graja, and E. Sopa
Physical Properties of the TCNQ Salts with 1-Methyl-3-Alkylimidazolium Derivatives
(72.20 and 78.30; S12)
(b) Vol. **140**, No. 2, p. 545—558 (1987).

W. Osak and K. Tkacz
Long-Lasting Relaxation Currents in TGS
(77.80; S12)
(a) Vol. **100**, No. 2, p. 667—672 (1987).

A. J. Pal, G. B. Talapatra, and T. N. Misra
Concentration Effect on Delayed Emissions: 1,4 Dibromonaphthalene Host–Pyrene Guest Mixed Crystal System
(78.55; 71.35; S12)
(b) Vol. **141**, No. 1, p. 351—360 (1987).

W. Osak and K. Tkacz
An Improved Way of Obtaining *I–U* Characteristics of Ferroelectrics
(77.80; S12)
(a) Vol. **101**, No. 1, p. 263—266 (1987).

H. Kresse, W. Weissflog, C. Nagabhushan, and R. Shashidhar
Dipole–Dipole Interactions in Swallow-Tailed Liquid Crystalline Mixtures
(77.20; S12)
(a) Vol. **101**, No. 1, p. K77—K79 (1987).

T. N. Sairam and S. V. Bhat
EPR Study of Mn^{2+} Doped Dicalcium Barium Propionate Single Crystals under Hydrostatic Pressure
(76.30; 64.70; S12)
(b) Vol. **141**, No. 2, p. K133—K138 (1987).

C. Sugiura, M. Kitamura, and S. Muramatsu
Palladium L_{III} X-Ray Absorptions-Edge Spectra of $Pd(ac)_2$, $Pd(acac)_2$, and $Pd_2(dba)_3 \cdot CHCl_3$
(78.70; S12)
(b) Vol. **141**, No. 2, p. K173—K177 (1987).

P. L. Muthal, B. T. Deshmukh, and S. V. Moharil
Thermoluminescence in Extended Chain Crystals of Polyethylene
(78.60; S12)
(a) Vol. **101**, No. 2, p. 583—587 (1987).

ZHENGMIN SUN, HUAQIN WANG, DIRONG QIU, and LIANGYU WANG
X-Ray Diffraction Investigations in the Disc-Like Mesophase of Three Compounds of Rufigallol-hexa-n-alkanoates
(61.30; S12)
(a) Vol. **101**, No. 2, p. K93—K96 (1987).

L. Q. MINH, T. CHOT, N. N. DINH, N. N. XUAN, N. T. BINH, D. M. PHUOC, and N. T. BINH
A New Organic Photoconductor in the Near Infrared Region of Diode Lasers
(73.60 and 78.65; S12)
(a) Vol. **101**, No. 2, p. K143—K146 (1987).

S. I. GAIDAI, K. P. MELETOV, and M. F. SHCHANOV
Quasi-Resonance in Hydrostatically Compressed Isotopic Mixed Crystals of Deuteronaphthalene
(71.35 and 71.55; 78.50; 78.55; S12)
(b) Vol. **142**, No. 1, p. 211—220 (1987).

S. N. KONYAEV, A. I. SHUSHIN, L. I. KOLESHNIKOVA, M. M. TRIBEL, and E. L. FRANKEVICH
Fluorescence-Detected Low Field Resonances in T–T Annihilation in the A-TCNB Single Crystal. Lineshape and Dynamic Parameters
(71.35; 76.90; 78.55; S12)
(b) Vol. **142**, No. 2, p. 461—471 (1987).

T. GOWOREK
Defects in Molecular Solids. Positron Studies
(61.70 and 78.70; S12)
(a) Vol. **102**, No. 2, p. 511—526 (1987).

A. V. GOLDANSKII, V. A. ONISHUK, and V. P. SHANTAROVICH
Some Principles of the Studies of Positron Annihilation in Polymer Systems
(61.40 and 78.70; S12)
(a) Vol. **102**, No. 2, p. 559—564 (1987).

P. ARIFOV, S. VASSERMAN, and S. TISHIN
Study of Elastomer Structure Microdefects by Positron Annihilation
(61.40 and 78.70; S12)
(a) Vol. **102**, No. 2, p. 565—570 (1987).

F. H. HSU, B. G. WEN, J. F. R. KUCK, and N. T. YU
Positron Annihilation Studies of Age-Induced Changes in Animal Lenses
(78.70; S12)
(a) Vol. **102**, No. 2, p. 571—575 (1987).

T. ASAJI, M. TAYA, and D. NAKAMURA
Pyroelectricity of Meta-Substituted Nitrobenzenes m-$XC_6H_4(NO_2)$ ($X = NH_2$, Br, Cl)
(77.70; S12)
(a) Vol. **102**, No. 2, p. 815—818 (1987).

S. S. Bawa, A. M. Biradar, and S. Chandra
Molecular Reorientation Process and Non-Linear Polarization Reversal Current in Helicoidal Ferroelectric Liquid Crystal DOBAMBC
(77.80; S12)
(a) Vol. **102**, No. 2, p. 829—834 (1987).

A. Kuczkowski, J. Liwo, and M. Chybicki
Controllable Modification of Electrical Conductivity of Polymers by Means of Ion Beam Irradiation
(72.80; S12)
(a) Vol. **102**, No. 2, p. K113—K116 (1987).

P. Wünsche, R. Kunze, and M. Wäsche
Activation Energy Distribution in Poly-(Methylmethacrylate–Methacrylic Acid) Estimated by EPR and Thermoluminescence
(65.90; 76.30; 78.60; S12)
(a) Vol. **103**, No. 1, p. 145—152 (1987).

B. Kumar
Carrier Drift Mobilities in Some Organic Solids. A Search for Organic Photovoltaic Materials. II. Charge Transport Calculations in s-Tetrazine and 4-Phenylazoazobenzene
(72.40 and 72.80; 71.25; S12)
(b) Vol. **143**, No. 2, p. 663—673 (1987).

C. Medrano P., P. Günter, and H. Arend
Noncentrosymmetry Observed in $CaC_4H_4O_6 \cdot 4\ H_2O$ Crystals by Nonlinear Optical Measurements
(78.20; S12)
(b) Vol. **143**, No. 2, p. 749—754 (1987).

H. Puff and H. W. Stritwolf
Optical Absorption of the Soliton Lattice
(78.20; 71.45; S12)
(b) Vol. **143**, No. 2, p. 765—775 (1987).

M. A. Gaffar, M. M. Mebed, and A. Abu El-Fadl
The Specific Heat of Pure and Doped Triglycine Sulphate Single Crystals
(65.40 and 77.80; S12)
(a) Vol. **103**, No. 2, p. 459—466 (1987).

K. Ikezaki
Thermally Stimulated Currents in FEP Films with Sputtered Al Electrodes
(73.60; S12)
(a) Vol. **103**, No. 2, p. 481—489 (1987).

F. Y. M. Al-Eithan, M. C. Al-Edani, and A. K. Abass
Optical Band Gap of Triglycine Selenate (TGSe) Single Crystals
(78.20; S12)
(a) Vol. **103**, No. 2, p. 571—574 (1987).

S12.1. Quasi-one-dimensional systems

T. P. EGGARTER and A. TROPER
A Simple Cluster Improvement to the Coherent Potential Approximation
(71.10; 71.25; S4; S12.1)
(b) Vol. **140**, No. 1, p. 127—133 (1987).

L. FIRLEJ, A. GRAJA, A. RAJCHEL, K. WOŹNIAK, and T. M. KRYGOWSKI
The 7,7′, 8,8′-Tetracyano-p-Quinodimethanide Salt of the N-n Propyl Quinoxalinium Cation — Its Structure, Conductivity, and Triplet Spin Excitons
(61.50 and 72.80; 76.30; S12.1)
(b) Vol. **140**, No. 2, p. 437—447 (1987).

Z. IVIĆ, M. SATARIĆ, Z. SHEMSEDINI, and R. ŽAKULA
The Role of Supersonic Davydov Solitons for Energy Exchange between Two Joined Molecular Chains
(71.45; S12.1)
(b) Vol. **140**, No. 2, p. 467—476 (1987).
Erratum
(b) Vol. **143**, No. 2, p. 806 (1987).

V. ŽELEZNÝ, J. PETZELT, and R. SWIETLIK
Infrared Properties of $TeEA(TCNQ)_2$
(78.30; 63.20; 71.38; S12.1)
(b) Vol. **140**, No. 2, p. 595—604 (1987).

A. V. SHERMAN
Energy and Absorption Spectra of One-Dimensional Exciton–Phonon System
(71.35 and 78.40; S12.1)
(b) Vol. **141**, No. 1, p. 151—161 (1987).

M. V. SIMONYAN, E. A. KAFADARYAN, M. CH. NURIJANYAN, A. K. PETROSYAN, and E. G. SHAROYAN
ESR, Optical Absorption and Reflection Spectra of Monoclinic and Triclinic Modifications of Lead Phthalocyanine
(76.30 and 78.20; 78.30; S12.1)
(a) Vol. **101**, No. 1, p. 143—149 (1987).

A. J. DANN, M. R. FAHY, C. JEYNES and M. R. WILLIS
Electrical Properties of Ion Implanted PcAlF Films
(73.60; S12.1)
(a) Vol. **101**, No. 1, p. K45—K48 (1987).

B. V. COSTA and A. S. T. PIRES
Properties of Kinks in a Driven Damped Anisotropic Magnetic Chain
(71.45 and 75.50; S12.1)
(b) Vol. **142**, No. 1, p. 117—123 (1987).

Author Index

Attention

To find a paper of your interest please choose first the special entry of the subject index and then use the volume number and page.

International Classification System for Physics*)

60. Condensed matter: structure, mechanical and thermal properties
    61. Structure of liquids and solids; crystallography
    (see also 68.20. Solid surface structure, 71. Electron states)
    62. Mechanical and acoustical properties of condensed matter
    (see also 61.70. Defects in crystals, 68.30. Surfaces and interfaces)
    63. Lattice dynamics and crystal statistics
    (see also 65. Thermal properties, 66.70. Thermal conduction, 68.30. Dynamics of surface and interface vibrations, 78.30. Infrared and Raman spectra)
    64. Equations of state, phase equilibria, and phase transitions
    65. Thermal properties of condensed matter
    (see also 63. Lattice dynamics; for thermodynamic properties of quantum fluids, see 67.40; for thermal properties of solid helium, see 67.80)
    66. Transport properties of condensed matter (nonelectronic)
    67. Quantum fluids and solids; liquid and solid helium
    68. Surfaces and interfaces; thin films and whiskers
    (for impact phenomena, see 79; for crystal growth, see 61.50)
70. Condensed matter: electronic structure; electrical, magnetic, and optical properties
    71. Electronic states
    (see also 63. Lattice dynamics, 73. Electronic structure and electrical properties of surfaces, interfaces, and thin films)
    72. Electronic transport in condensed matter
    (for surfaces, interfaces, and thin films, see 73)
    73. Electronic and electrical properties of surfaces, interfaces, and thin films
    74. Superconductivity
    75. Magnetic properties and materials
    76. Magnetic resonances and relaxation in condensed matter; Mössbauer effect
    77. Dielectric properties and materials
    (for conductivity phenomena, see 72.20 and 72.80)
    78. Optical properties and condensed-matter spectroscopy and other interactions of matter with particles and radiation
    (for phonon spectra, see 63)
    79. Electron and ion emission by liquids and solids; impact phenomena
    85. Devices**)

---

*) Excerpt; reproduced with permission of International Council for Scientific and Technical Information (ICSTI).

**) Outside the ICSTI Classification for Physics.

(The Substance Classification is given on cover three)

Manuscripts and letters for physica status solidi (b) — basic research and physica status solidi (a) — applied research

should be sent to

Leipziger Straße 3 4, P. B. 1233, DDR-1086 Berlin

or

Kunst und Wissen, Erich Bieber OHG, Postfach 330104, D-1000 Berlin 33.

The object of "physica status solidi (b)" and "physica status solidi (a)" is to publish rapidly important and topical results concerning fundamental research in the field of solid state physics and aspects of application and experimental techniques of fundamental solid state research, respectively. The issues of "physica status solidi (b)" and "physica status solidi (a)" will be published on the 1st and the 16th day of each month, respectively. Either series will contain the following categories:

1. Review articles written by arrangement with the editors. They will be remunerated.
2. Original papers selected from those which have not been submitted for publication elsewhere, after being reviewed. (The latest date for acceptance of contributions is the 8th and the 23rd day of each month for "physica status solidi (b)" and "physica status solidi (a)", respectively, 50 days before the delivery of the journal.)
3. Short notes published in the off-set part of the journal. They are, if necessary after being reviewed, selected by editorial staff. (The latest date for acceptance of manuscripts for this part is the 12th and the 27th day of each month for "physica status solidi (b)" and "physica status solidi (a)", respectively, 16 days before the delivery of the journal.)
4. Pre-printed titles of papers to be published in the next issue of both series.

Editorial Information

Manuscripts should be submitted preferentially in English. Besides, papers in German, French, and Russian are admitted. The final decision on the incorporation of the papers in the series (b) or (a) is taken by the editorial staff after consultation of the referee. Papers on liquids are only published in physica status solidi if the problems treated are also relevant for the solid state. Authors are asked to observe the following rules:

1. Manuscripts have to be written as consisely as possible. At most, 20 and 4 pages (28 lines per page; abstracts, tables, references, figures, and captions of figures included, where 3 figures are counted as about one page) are admitted for original papers and short notes, respectively. All manuscripts should be typewritten on one side only. Two copies are required. Badly arranged manuscripts are returned.
2. The author's name, initials, and institution should be given.
3. One abstract of about ten lines in English and one in German, French, or Russian are required for original papers.
4. Because of lack of space conclusions (if any) must not have the character of a summary (i.e. substantially a repetition of the contents of the abstract).
5. Formulae should not be written too small and not with pencil. Separate lines for formulae are desirable. Indistinct writing will lead to a return of the manuscript.
6. Drawings should be enclosed on separate sheets of paper, not exceeding the maximum and minimum sizes of 20×30 and $10 \times 15\ cm^2$, respectively. Photographs should be submitted, on white paper, only for review articles and original papers. Duplicates are required.
7. Captions for the figures should be submitted on separate sheets of paper, and special instructions with regard to the size of illustrations should be given on these.
8. References should be numbered in order of appearance and should be given, on a separate sheet, in the form used in physica status solidi.
9. SI-units should be used.
10. Each paper must contain a proposal for the subject classification according to the classification scheme appearing in the January issues of each annual set.

The authors will be sent galley proofs of review articles and original papers. Proof reading of short notes and pre-printed titles will be done by the editorial staff.

For each original paper and short note a total of 75 reprints is available free of charge. The number of reprints free of charge per review article is 25. Requests for additional reprints (against payment) should be given together with the submitted manuscripts. **The deadline for orders of additional reprints is the 15th and the 30th of each month** for "physica status solidi (b)" and "physica status solidi (a)", respectively, two weeks before the delivery of the journal.

Bei Fragen zur Produktsicherheit wenden Sie sich bitte an:
If you have any questions regarding product safety,
please contact:

Walter de Gruyter GmbH
Genthiner Straße 13
10785 Berlin
productsafety@degruyterbrill.com